THE LEGO® LIGHTING BOOK

LIGHT YOUR LEGO® MODELS!

BRIAN M. WILLIAMS

no starch
press

SAN FRANCISCO

Printed in China

First printing

27 26 25 24 23   1 2 3 4 5

ISBN-13: 978-1-7185-0084-6 (print)
ISBN-13: 978-1-7185-0085-3 (ebook)

Publisher: William Pollock
Managing Editor: Jill Franklin
Production Manager: Sabrina Plomitallo-González
Developmental Editor: Nathan Heidelberger
Production Editors: Sabrina Plomitallo-González and Miles Bond
Cover Design: Monica Kamsvaag
Cover Model: *Kent Theatre*, designed and built by Jarren Harkema
Interior Design and Composition: Maureen Forys, Happenstance Type-O-Rama
Copyeditor: Rachel Monaghan
Proofreader: Paula L. Fleming

For information on distribution, bulk sales, corporate sales, or translations, please contact No Starch Press, Inc. directly at info@nostarch.com or:

No Starch Press, Inc.
245 8th Street, San Francisco, CA 94103
phone: 1.415.863.9900
www.nostarch.com

Library of Congress Cataloging-in-Publication Data

Names: Williams, Brian M. (LEGO artist), author.
Title: The LEGO lighting book / Brian M. Williams.
Description: San Francisco : No Starch Press, [2023] | Includes index. |
Identifiers: LCCN 2022059259 (print) | LCCN 2022059260 (ebook) | ISBN
  9781718500846 (print) | ISBN 9781718500853 (ebook)
Subjects: LCSH: LEGO toys. | Lighting, Architectural and decorative. | LED
  lighting. | Models and modelmaking.
Classification: LCC TS2301.T7 W48 2023 (print) | LCC TS2301.T7 (ebook) |
  DDC 688.7/25--dc23/eng/20230130
LC record available at https://lccn.loc.gov/2022059259
LC ebook record available at https://lccn.loc.gov/2022059260

The endpapers depict various types of lamps made entirely from LEGO bricks. Can you find (1) the first practical incandescent lamp that Thomas Edison used in his first public demonstration at Menlo Park, NJ, on New Year's Eve 1879, which is currently at the Smithsonian: *https://www.si.edu/newsdesk/snapshot/edison-light-bulb*; (2) an improved Edison lamp patented on January 27, 1880, currently at the Franklin Institute: *https://www.fi.edu/history-resources/edisons-lightbulb*; (3) a replica of Edison's first light bulb, currently at the Thomas Edison National Historic Park: *https://www.nps.gov/edis/learn/kidsyouth/the-electric-light-system-phonograph-motion-pictures.htm*; (4) the Centennial Light—the world's longest-lasting light bulb, which has been burning for 120 years—in Livermore, CA: *https://www.centennialbulb.org*; and (5) the first-mass produced visible light LED, made by Monsanto in 1968: *http://www.lamptech.co.uk/Spec%20Sheets/LED%20Monsanto%20MV1.htm?*

This book is dedicated to the Lord of the universe,
who hath given to us both light and shadow, sight
to appreciate His world's beauty through them,
and to me a talent for sharing them.

ABOUT THE AUTHOR

Brian M. Williams is a longtime member of the LEGO community and a coordinator for Brickworld Chicago, where he has twice received the World of Lights award. He has been combining his passions for LEGO and lighting since he was six years old when he wired LEDs into his first LEGO spaceship MOC. As an adult, he has 28 years of experience building scale model railroads, military miniatures, and LEGO creations.

CONTENTS

PREFACE

As a boy, I was fascinated with LEGO and electronics. As an adult, I've combined those interests to build—and artfully light—my most satisfying custom LEGO creations. Now I'm pleased to share with you the lessons I've learned along the way.

We live in a golden age of sorts, when advancing LED technology is not only transforming entire industries and reducing the world's energy consumption but also giving LEGO enthusiasts a fascinating new way to enjoy their creations: in the dark. LEDs are tiny yet incredibly bright. They have rich, saturated colors, generate little heat, and are affordable enough to light up entire LEGO cities. And to light those cities, an industry of third-party manufacturers has sprung up to sell custom LED lighting products especially made for use with LEGO.

As you'll learn in this book, there's a world of difference between just installing lights functionally and sculpting with light creatively. It's the difference between painting a room and painting the *Mona Lisa*. Just as building with LEGO offers infinite creative possibilities, so, too, does lighting LEGO creations. This book explores those creative possibilities, teaches proven techniques, and showcases amazing examples of those techniques in use. It shares universally applicable lighting principles developed in fields from art to architecture, as well as those discovered in the LEGO fan community.

WHO IS THIS BOOK FOR?

There are many LEGO enthusiasts who will want to read this book. Some are longtime LEGO builders, like myself, who are part of the Maker culture—DIY types familiar with electronics. Others are more into collecting LEGO sets or are new to the hobby and may be less familiar with electronics. What they all have in common is that they're fascinated by tiny LEDs and want to know how to light up their own miniature LEGO worlds. Sound familiar? Then this book is for you.

WHAT'S IN THE BOOK

This book is organized topically. After setting the stage in Chapters 1 and 2, each remaining chapter introduces techniques appropriate to a popular category of LEGO lighting and showcases a gallery of inspiring examples. There is, of course, overlap, with some techniques shared across categories, but the book's chapter divisions will help you quickly locate the techniques most applicable to the type of creation you want to light.

Here's what you'll find in the book:

CHAPTER 1: LIGHT BECOMES ART

Introducing the magic of LEGO lighting through a trip to Brickworld Chicago, home of the first LEGO lighting competition

CHAPTER 2: LIGHTING SOLUTIONS

Surveying the many lighting options available to LEGO creators, including official lighting elements from the LEGO Group, third-party products made especially for use with LEGO bricks, generic commercial products like hobby lights and strip lights, and DIY solutions

CHAPTER 3: INTERIOR LIGHTING

Lighting rooms inside buildings, including general illumination as well as specialty lighting fixtures

CHAPTER 4: EXTERIOR LIGHTING

Lighting building exteriors, including window lighting, landscape lighting, and streetlights, and managing wiring for single or multiple buildings

CHAPTER 5: LIGHTING WITH CHARACTER

Lighting up figures and figure accessories like light swords, flashlights, control panels, and heads-up displays

CHAPTER 6: VEHICLE LIGHTING   Adding

lights to cars, trucks, aircraft, spaceships, and more

CHAPTER 7: SHADOW ART   Creatively

using bricks to throw shadows

CHAPTER 8: GLOWING BRICKS   Using

trans-color bricks, glow-in-the-dark bricks, and bricks that glow under ultraviolet light to create sculptures lit from the inside

CHAPTER 9: DYNAMIC LIGHTING   Using

lighting effect controllers to animate lights on emergency vehicles, spacecraft, and more

Much of the book's terminology comes from the LEGO fan community or one of the discussed fields, such as electronics or cinematography. If a term is unfamiliar, please consult the extensive glossary at the back of the book. The LEGO element names, numbers, and colors referenced throughout match the ones used on BrickLink (*https://www.bricklink.com*), the world's largest online marketplace to buy and sell LEGO parts.

ABOUT THE PROJECTS

Each chapter features step-by-step building instructions for a hands-on project to apply what you've learned. You can purchase the necessary LEGO parts online through Brick-Link. Visit the book's web page at *https://nostarch.com/lego-lighting-book* to download parts lists and check for any updates.

For convenience, lighting manufacturers Brickstuff (*https://www.brickstuff.com*), LifeLites (*https://www.lifelites.com*), and Brick Loot (*https://www.brickloot.com*) sell dedicated kits with all the lighting components you need to build this book's projects.

The type and placement of the LEDs are specified in the instructions, but the choice of power source (battery, USB, and so on) is yours. Some models provide space to incorporate a small coin-cell battery into the build, if desired.

PLAY WELL

Like a good LEGO build, this book has been a labor of love. I hope that you have as much fun reading it as I've had writing it. But above all, I hope my book motivates you to experiment with lighting your own creations. There's a lot of information to unpack in these pages. Some of it may be useful today, other parts later, so refer back frequently as you begin your new LEGO projects. Because also like with a good LEGO build, you'll discover new things each time you come back.

Considering the wide range of creations showcased in this book, you might be inclined to think that it's all been done already. Yet each year LEGO fans find new, creative ways to marry lights and bricks. You may become inspired by reading this book to develop as-yet-unknown techniques that will in turn illuminate others. After all, human ingenuity and creativity, like the possibilities of the LEGO brick itself, are boundless.

1 LIGHT BECOMES ART

To love beauty is to see light.

—VICTOR HUGO

Humankind has been fascinated by the nature of light since we first looked up and contemplated the sun. Light is functional. It informs our sight, directs our senses, and enables most of life's activities. It's so familiar that we take it for granted. Yet at times light rises above the utilitarian to entertain, mystify, and even inspire. The glimmer of sun dancing off the lake on a warm summer morning, the cool glow of moonlight at night, and the warm waves of color that blanket clouds at sunset . . . light paints these images using an emotional language innately understood by every heart. And it's in our perception of these emotions that light takes on a whole new reality. It transcends function and becomes art.

Henley Street Bridge (Peter Campbell), winner of the Brickworld 2018 World of Lights award

There's a branch of the visual arts called *light art* in which light itself becomes the artistic medium. Examples range from glowing acrylic sculptures to the use of powerful lamps that project images onto large buildings. Light art can be as small as a neon sign or as large as the light shows created for the Eiffel Tower, involving thousands of flashing lights. LEGO, with its flexible array of trans-colored bricks and lights, lends itself to artistic experimentation with light. AFOLs (Adult Fans of LEGO) took notice of these qualities, and experiment they did.

A WORLD OF LIGHTS

Perhaps the best way to appreciate the appeal of LEGO lighting is to visit an annual LEGO fan event called the Brickworld World of Lights. Brickworld Chicago is one of the largest LEGO fan conventions in North America. Back in 2008, Brickworld co-founder Adam Reed Tucker wanted a new activity to round out the evening schedule. He thought about asking attendees to light their MOCs (*My Own Creation*, or *MOC*—pronounced "mock"—is the term AFOLs use to describe their original designs) for a new annual event where the

The first World of Lights at Brickworld Chicago 2009

World of Lights at Brickworld Chicago 2011

convention hall lights would be turned down. After discussing with his management partner, Bryan Bonahoom, Adam launched the World of Lights challenge, which premiered at Brickworld Chicago 2009. LEGO lighting enthusiast Rob Hendrix helped with judging and built a unique lighting award trophy.

No one knew what to expect that first year. Would anyone participate? Would the attendees, many being LEGO purists, use only official LEGO lighting elements? As the lights went down that first night, it became clear that lighting was something the AFOLs truly embraced. Dozens of creations were lit, from dragons to diners, emergency vehicles to spacecraft, town streets to movie scenes. And they weren't just lit with official LEGO elements but with ordinary light bulbs, battery-powered clip-on lights, and hand-soldered LEDs, too. Some lighting installations were as simple as placing a light bulb inside a large translucent LEGO cylinder. The more complex installations involved snaking long wires and lights carefully through buildings. I was among those present that night, where I was honored to receive the first World of Lights award for my shadow box dioramas of movie scenes.

Walking around the World of Lights is surreal. Beforehand everyone is busily making last-minute adjustments and testing out their mechanisms. Then the room lights go down, and the crowd cheers with enthusiastic applause as a thousand tiny lights begin to glimmer in the vast darkness. After a minute your eyes adjust, and the MOCs come to life, taking on a different character than during lighted hours. Everyone circulates around the room, studying the carefully lit scenes and enjoying the spectacle. Some MOCs that didn't draw a lot of attention in the light are suddenly thronged with fans. Music plays, creating a festive, rock concert–like vibe. Everyone is relaxing in this collective Lilliputian fantasyland that goes on until early in the morning.

The Dragon's Wrath (Cornbuilder), Brickworld 2018

Octan Park (Ryan Degener), Brickworld 2017

Venice Library (author), Brickworld 2009

Deathstar Disco (John Wolfe), Brickworld 2011

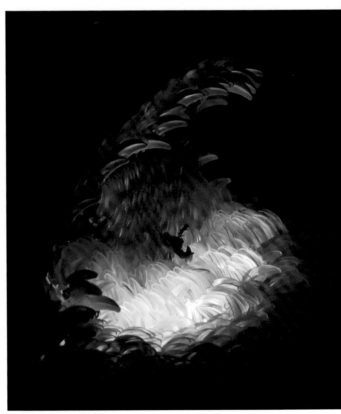

Kubo and the Two Strings (Amanda Feuk and Jesse Mohler), Brickworld 2018

Since 2009, the World of Lights has grown, with more and more attendees contributing an ever-increasing range of lighted MOCs each year. Meanwhile, the Brickworld World of Lights has inspired other LEGO fan conventions, like BrickFair, to hold similar lights-out events. Between Brickworld and BrickFair over the years, there have been 28 LEGO lighting competition winners to date. Their discoveries and techniques form many of the examples in this book.

The Tour Continues (Marc Hendricks), Brickworld 2015

Tree of Worlds (Eurobricks), Brickworld Chicago 2018

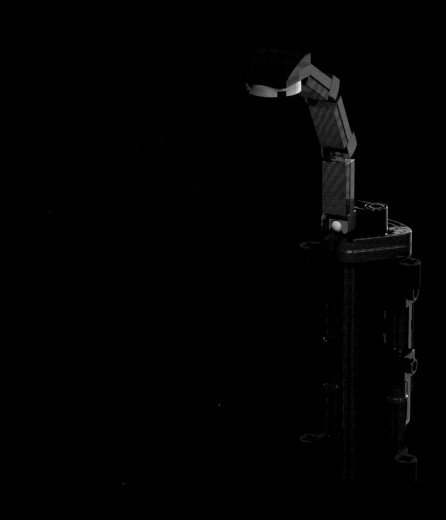

Figure 1-1: Project 1: Book light

PROJECT 1: BUILD YOUR OWN BOOK LIGHT

Now it's time to get hands-on and build your own LED-powered book light (see Figure 1-1). This way, you never have to put this book down, even in the dark! Follow the building instructions using your own bricks, plus a few electronic components. (For an itemized list of the parts needed, please check *https:// nostarch.com/lego-lighting-book*.)

Requirements
- One pico-style LED, such as a Brickstuff Pico LED or equivalent (see Chapter 2 for more information). This type of LED is small enough to fit under a 1×1 round LEGO plate and has wires thin enough to fit between bricks. You can substitute a LEGO battery-powered Light Brick (BrickLink #54930c02) for the pico LED if you like.

- One adapter board, as needed, to connect the LED to the power source, such as the Brickstuff BRANCH19 adapter or equivalent.

- One power source compatible with the LED used. This project provides space on the back of the booklight to attach a small coin-cell battery pack if desired, such as a Brickstuff Coin Cell Battery Pack.

- Double-sided tape, as needed, to attach the battery pack.

4x
6541
Black

1x
2780
Black

1x
6536
Black

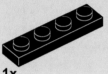

1x
3710
Black

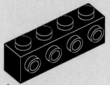

1x
30414
Black

1x
14704
Dark Bluish Gray

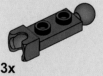

3x
14419
Dark Bluish Gray

1x
2736
Light Bluish Gray

1x
4032
White

2x
3713
Red

3x
3069b
Red

1x
15068
Red

2x
32249
Red

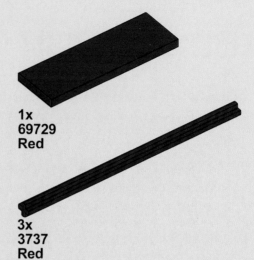

1x
69729
Red

3x
3737
Red

1x
2654
Trans-Clear

4x
45590
Rubber Black

1

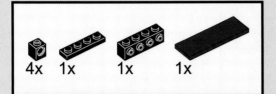

4x 1x 1x 1x

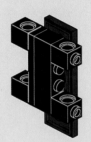

2

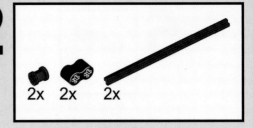

2x 2x 2x

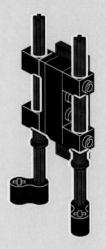

3

1x 2x 2x 1x

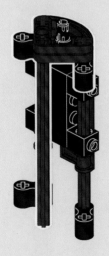

4

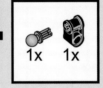

1x 1x

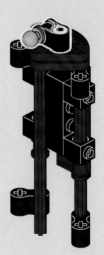

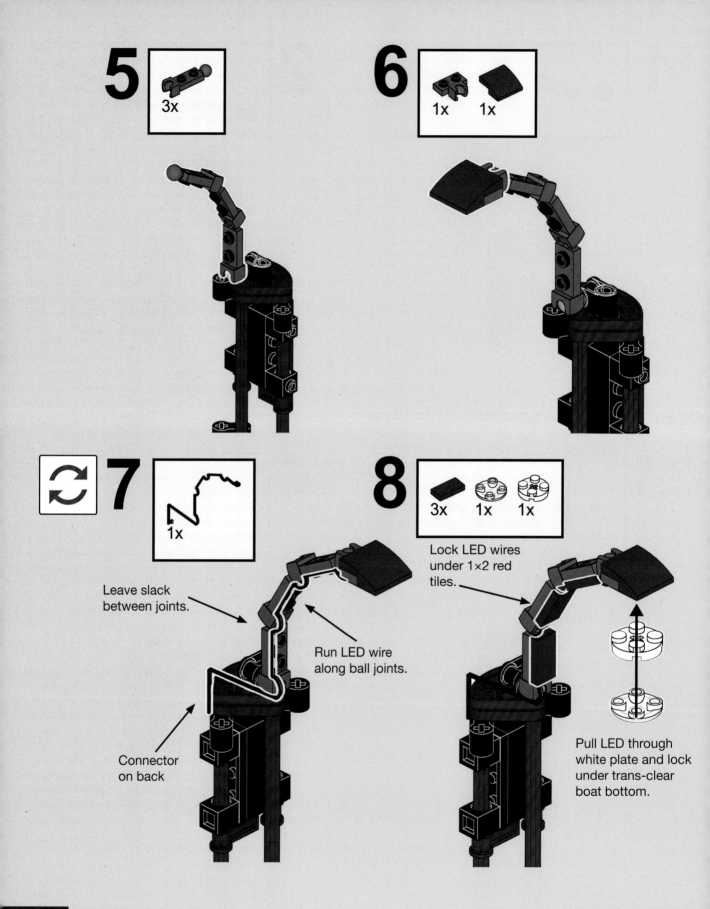

5 3x

6 1x 1x

7 1x

Leave slack between joints.

Run LED wire along ball joints.

Connector on back

8 3x 1x 1x

Lock LED wires under 1×2 red tiles.

Pull LED through white plate and lock under trans-clear boat bottom.

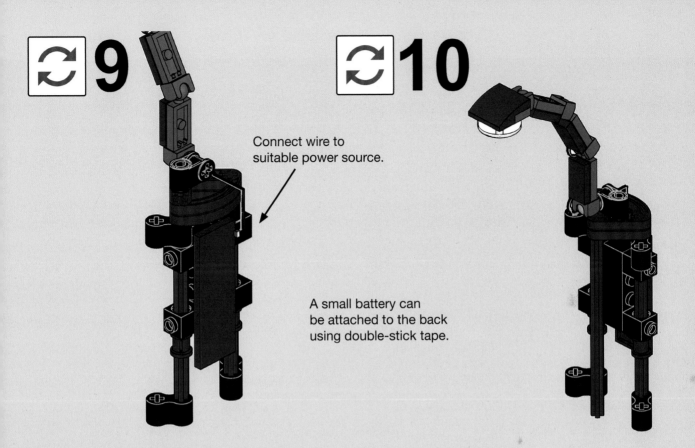

9

Connect wire to
suitable power source.

A small battery can
be attached to the back
using double-stick tape.

10

2 LIGHTING SOLUTIONS

Light makes photography. Embrace light. Admire it. Love it. But above all, know light. Know it for all you are worth, and you will know the key to photography.

—GEORGE EASTMAN, PHOTOGRAPHY PIONEER AND ENTREPRENEUR

Our passion for recording light and shadows, commonly known as photography, is at once both art and science. While the art has much to teach us about how to use light evocatively, the science captures the evolution of artificial light sources, culminating with today's ubiquitous *light-emitting diodes (LEDs)*, which are far more efficient and long-lasting than earlier incandescent, fluorescent, and other lighting technologies. When LEDs were first

150 in One Lighting Project Kit (author). All the equipment is made from bricks and includes some Easter eggs for LEGO fans. Can you find them? For the answer, turn to page 35.

commercialized in the 1960s, they were the domain of electronics tinkerers. Today, LEDs come in all sizes and colors, and they've revolutionized commercial lighting markets, including toys.

Thanks to this LED revolution, there are now many options to choose from when lighting your LEGO MOCs. If you prefer to use only official LEGO-made elements, you can draw on any of the official lighting systems that LEGO has produced over the last 60 years. If you want the brightest lights with the fastest installation, you can pick from several third-party lighting companies that make LED lighting systems specifically for use with LEGO bricks. Or, if economy is your thing, you can look into commercial lighting solutions or adopt a do-it-yourself (DIY) approach and solder your own lighting components together. This chapter provides an overview of these various options and introduces many of the concepts and terms that will be used throughout the rest of this book.

OFFICIAL LEGO LIGHTING SYSTEMS

The LEGO Group has produced six different lighting/electrical systems over the years: 4.5 V, 12 V, 9 V, Battery, Power Functions, and Powered Up. Indeed, lighting has been a part of the LEGO system almost from the start. The first LEGO lighting element appeared in 1958, the same year the LEGO brick achieved its final form, with tubes underneath. The 4.5 V lighting element, shown in Figure 2-1, was introduced in set #245-1, Lighting Device Pack.

This section profiles the features of each of LEGO's six lighting systems. If you're a LEGO purist and want to use only official LEGO elements in your creations, then read carefully about each lighting system. But even if you aren't a purist, some older LEGO lighting elements are unique and may still have a place in your brick-lighting toolkit. Note, however, that most of LEGO's 4.5 V, 12 V, and 9 V lights use

Figure 2-1: The first LEGO light brick, from set #245-1, Lighting Device Pack (1958)

incandescent bulbs, which draw higher current and generate more heat than LEDs. This limits the number of bulbs that you can power simultaneously.

Figure 2-2: Elements from the 4.5 V lighting system

4.5 V LIGHTS

The first LEGO light brick from 1958, #x456c01, featured a 4.5 V incandescent bulb housed in a transparent (*trans-clear* in BrickLink parlance) 2×4 brick with two holes on the back (see Figure 2-1). These light bricks could be connected singly, in parallel, or in series using the wires included in the Lighting Device Pack set, which had plugs on each end. LEGO didn't make a battery box initially, so each plug had a split that allowed it to attach to the thin metal terminals of a 4.5 V type 3LR12 battery. A special 1×2 brick was also included in the set. It featured a small rectangular notch on the bottom, permitting the wires to be passed underneath the brick.

Over time, LEGO introduced more 4.5 V components, including a 4.5 V battery box, new wires and plugs, and 2×2 light bricks molded in yellow and red (see Figure 2-2, top center). Consistent with the overall LEGO tenet of interchangeability, the 4.5 V elements were originally conceived as accessories that could be purchased separately to add light to any set or MOC. The wires came in multiple lengths and could be daisy-chained to extend their reach or connect multiple lights together. Accessory bricks were available in several transparent colors, some printed with words like *Taxi* or *Station*, to place in front of the lights. These accessory bricks, some of which are shown in Figure 2-2, can all still be used with later lighting systems to create signs and effects.

12 V LIGHTS

In 1980, LEGO introduced the gray-rail 12 V train line, marking the first time that lighting bricks were included in actual building sets. The new 12 V light bricks (Figure 2-3, top row) looked similar to the older 4.5 V 2×2 light bricks, but were in white and black and had a third small hole on the back to differentiate them from the 4.5 V lights. The new 12 V wires added a small plastic plug in the center of their connectors, also to differentiate them from their 4.5 V counterparts. The new 12 V lights could be powered by a new #7864 12 V Transformer. The #4170 and #4171 1×6 Prism Brick (Figure 2-3, bottom right) was introduced to make headlights and can be used today with other lighting systems. The prisms channel light from the rear of the 1×6 brick to the holes in the front of the brick. Notably, the 12 V train line also marked the first appearance of LEDs in a LEGO product, with the #70022 Signal Light Brick (Figure 2-3, bottom left).

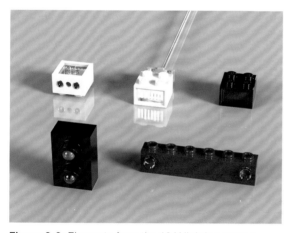

Figure 2-3: Elements from the 12 V lighting system

9 V LIGHTS

In 1986, LEGO introduced the 9 V system, which integrated lights, sounds, and motors with innovative new electrically conductive plates that could be built into creations. Battery boxes, wire cables, and a new 9 V wall-powered speed regulator (made to control train speed but easily used to power lights) were added, and all featured the same electrically conductive plates. The plates and

connectors could be rotated 90 degrees and still conduct electricity. The 9 V lights would stay lit if attached to the conductive plates in one direction, and they would flash on and off if connected in the opposite direction.

Some of the highlights from the 9 V system are shown in Figure 2-4, including some unique accessory bricks that are still highly desirable and useful today. For example, the #x85 parabolic reflector (bottom right) can be used to collimate light (make the light rays parallel, as does the reflector in a flashlight). The #2383 and #x905 light cabinets (top row) let you make cool backlit signs and control panels that can be lit by other lighting systems. New prism bricks were provided for train headlights (#2919/#2928; third row, left) and town traffic signs (#2500c01; bottom).

The 9 V system even provided several ways to animate lights for signs, signals, and sci-fi effects. For example, the unusual #6637 Fiber Optic element (top right) could mechanically cycle red light through eight different fiber optic strands. The #2840c01 Technic Control Center (not shown) was an electronic programmable controller that provided limited manual programming to make three sets of lights flash in different sequences. Or you could connect a LEGO MINDSTORMS robotics controller to provide even more sophisticated programming of three sets of lights. The ultimate official LEGO tool for lighting effects animation, however, was the #2954 LEGO Control Lab Serial Interface sold through LEGO's DACTA Educational Division. This

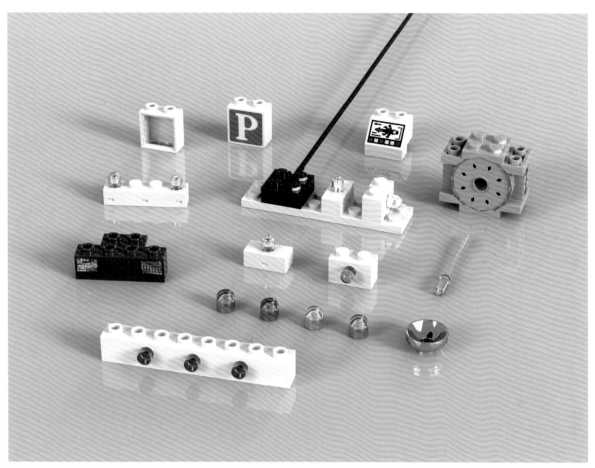

Figure 2-4: Elements from the 9 V lighting system

device permitted up to eight sets of lights to be fully programmed from a personal computer. It's easy to understand why the 9 V system holds special appeal for AFOLs even to this day.

BATTERY LIGHTS

LEGO began making self-contained, LED, battery-powered light and sound bricks in the late 1990s. These were large, single-use elements—for example, the #x239 Light & Sound Insectoid Stinger—to provide light and sound effects in space-themed sets. In 2006,

LEGO introduced the 2×3 battery-powered light brick (see Figure 2-5, right). Still in production today, these bricks are far smaller, more flexible, and more economical than earlier battery-powered lights. Inside is a red or yellow LED and two LR41 1.5 V batteries. They turn on with a push button on the rear of the brick.

These 2×3 light bricks are self-contained, affordable, and have appeared in almost 50 sets to date. They're portable and flexible enough to be used for headlights, rocket exhaust, chandeliers—anywhere that light is needed. The downside is that battery lights aren't the brightest and, of course, the batteries must be changed periodically.

POWER FUNCTIONS LIGHTS

In 2007, LEGO introduced the Power Functions (PF) system, which featured new four-wire cables. Two of these wires carried full 9 V current, and the other two carried variable power, enabling you to adjust the brightness of lights remotely via infrared controllers. However, the bulkier cables and larger, non-rotating 2×3 plate connectors were less flexible than the earlier 9 V connectors.

While several good PF-compatible motors and other accessories would eventually be released, the system ultimately produced only one light element: #8870 Light Unit (see Figure 2-5, left). It featured two LEDs permanently wired to a PF plug through a Y connector. Unfortunately, these LEDs were somewhat dim compared to those in other contemporary toys and third-party LED lighting systems.

POWERED UP LIGHTS

In 2018, LEGO introduced the current Powered Up system, which features new six-wire

Figure 2-5: Power Functions and Battery Lights. The Power Function lights (left) are perfect for car headlights. The Battery Lights (right) are flexible enough for any build.

cables with plug-in connectors. This system offers more sophisticated components that communicate with each other via the extra wires and can be controlled by software running on portable devices via a Bluetooth interface. Hence, the possibilities are intriguing. As of this writing, however, only one light has been introduced to the Powered Up system, the #88005 Powered Up Light. It's nearly identical to the older #8870 Power Functions Light Unit, except for the updated connector. Still, this new system holds great promise, and it will be interesting to see what new lighting elements might be introduced in the future.

DIY LIGHTING

Some LEGO enthusiasts opt to assemble their own LED lighting systems for their creations, producing custom components like the ones shown in Figure 2-6. This involves purchasing individual LEDs, wires, and so on and then soldering everything together yourself.

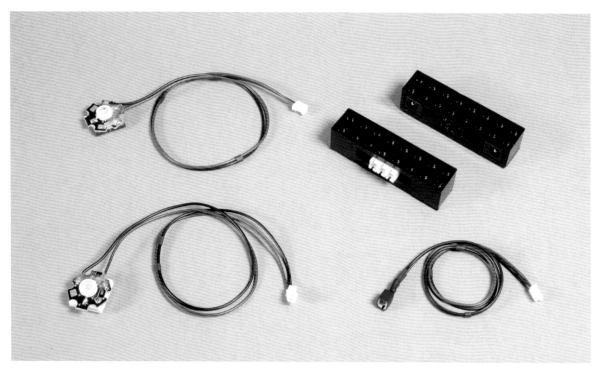

Figure 2-6: Examples of DIY lighting components, created by LEGO enthusiast John Wolfe: high-power UV LED (top left), high-power white LED (bottom left), extension wire (bottom right), and wall-wart power supply distribution hub (top right)

There are several advantages to the DIY approach. First, LEDs are widely available and inexpensive, as are the necessary supporting electronic components, such as wires and resistors. Thus, DIY is the lowest-cost approach to LEGO lighting. Additionally, doing it yourself provides the most control and flexibility and gives you access to the latest, brightest LEDs.

On the other hand, the DIY approach requires research and assembly work, both of which will cost you time. You also need basic math, soldering, and electronics skills, and in the end, the cost savings may be only a few dollars. Your time is worth something, too. That's why many people choose to use commercial or third-party lighting solutions and let someone else do the heavy lifting.

Even if you don't plan on making your own lighting components, please continue reading this section. It outlines the basics of LEDs, which also form the backbone of commercial and third-party lighting solutions, and presents concepts and terminology that will be used throughout the book.

LED BASICS

An LED, or light-emitting diode, is an electronic component that emits light when current flows through it. LEDs are more efficient and generate less heat than incandescent lamps. But unlike an incandescent lamp, which emits light when current flows through it in either direction, an LED emits light only when current flows in one direction.

Figure 2-7 shows the anatomy of a typical LED, with two wire leads and a plastic lens.

Inside the lens, connected to one lead, is a part called an *anvil*, which has a tiny reflective cup that holds a *semiconductor die*. The die is connected to a *post* by a hair-thin *bond wire*, and the post connects to the other lead. When the anvil is connected to the negative (–) side of a power source, and the post to the positive (+) side, then current flows through the semiconductor die, causing it to emit light. If the current is reversed, however, no light is produced.

There are many factors to consider when selecting LEDs. These include packaging style, size, voltage, current, viewing angle, brightness, and color. We'll briefly consider each in turn.

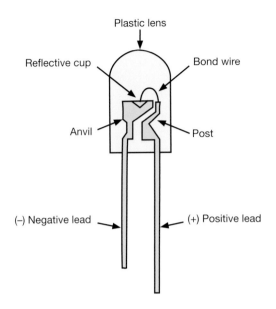

Figure 2-7: LED anatomy

Packaging Style and Size

You'll find three main types of LEDs on the market, named for the way they're packaged to be attached to a circuit board, and each type has its own range of sizes. *Through-hole LEDs* have long leads coming out of the lens, like the one shown in Figure 2-7. They're easy to work with and ideal for LEGO due to their size range. The 3 mm size fits nicely into the hole on a headlight brick, and the 5 mm size fits into the holes on LEGO Technic beams. *SMD (surface-mounted device) LEDs* are flat, so they can be surface-mount soldered. They come in smaller sizes, down to less than 1 mm, allowing them to fit through smaller holes in LEGO bricks. *COB (chip on board) LEDs* are also flat, but they're much larger

than SMDs. They combine many LEDs into a single unit for exceptional brightness, energy savings, and longer life.

Voltage

Individual LEDs typically operate between 1.2 V to 3.6 V, depending on the semiconductor chemistry used to generate the specific color. Red LEDs usually take the least voltage, while blue and white LEDs take the most. If the voltage supplied to an LED (the *supply voltage*) is higher than its operating voltage (the *forward voltage*), then a current-limiting resistor will need to be attached (see the section "The Four Cs of DIY Lighting" later in this chapter for more information on calculating resistor values).

Current

Most individual LEDs consume about 20 mA (milliamps) of current, though a COB package with multiple LEDs can draw 1 A (amp) or more. Be sure to check a seller's specifications to ensure you know the current draw, forward voltage, and supply voltage needed. This is important to consider when choosing a power supply for an LED installation.

ELECTRICAL CAPACITY

The laws of physics are universal and apply equally to all LED lighting manufacturers' products. The amount of available current from a power supply (its *capacity*) and the amount of current that LEDs consume are both measured in milliamps (mA). Each LED typically consumes around 20 mA. Thus, to determine the number of LEDs you can safely operate off a given power supply, simply divide the power supply's capacity in milliamps by 20. So, for a typical 1 A (1,000 mA) wall power supply, you can safely operate 50 LEDs (1,000 / 20 = 50).

Viewing Angle

Viewing angle refers to the size of area, measured in arc degrees, across which an LED will project light. Through-hole LEDs can feature 20-degree (narrowly focused) to 160-degree (widespread) viewing angles. SMD and COB LEDs typically have about a 120-degree viewing angle. Narrow viewing angles work best for headlights and spotlights, while wide viewing angles are best for backlighting windows, fire, and glowing objects. Separate attachment lenses are available for some LEDs to focus the light to narrower viewing angles.

Brightness

LED brightness, typically measured in millicandelas (mcd) or luminous flux (LUX), is affected by many factors: the size of the LED, the number of individual LEDs in a COB package, the viewing angle, the color, and even how the LED is powered, to name a few. Hence, it is difficult to determine relative brightness by specs alone. The best way to assess brightness is to examine individual LEDs in person and see how bright they appear to you. Note that to make interior spaces brighter, it's better to use several medium-brightness LEDs rather than a single high-brightness one.

Color

The color of light produced by an LED is a function of its *wavelength(s)*, measured in nanometers. For example, red is about 650 nm, green about 535 nm, and blue about 450 nm, as shown at the top of Figure 2-8. It's also possible to achieve a particular color by filtering the light from a white LED through translucent LEGO elements or another colored material. However, you'll achieve the greatest color saturation and brightness by using LEDs that emit the desired color.

INVISIBLE LIGHT

Some LEDs are designed to emit light beyond the range of human vision. This includes infrared LEDs, with wavelengths greater than 1,100 nm, and ultraviolet LEDs, with wavelengths less than 395 nm. Ultraviolet LEDs can be used to achieve a glowing effect with certain colored bricks (see Chapter 8 for more).

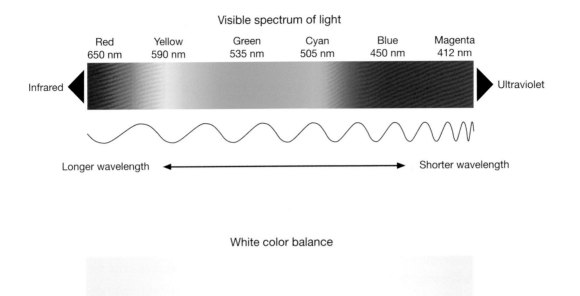

Visible spectrum of light

| Red 650 nm | Yellow 590 nm | Green 535 nm | Cyan 505 nm | Blue 450 nm | Magenta 412 nm |

Infrared — Ultraviolet

Longer wavelength ←——————→ Shorter wavelength

White color balance

Warm white
3,000 K

Daylight white
4,800 K

Cool white
7,000 K

Figure 2-8: LED colors

White LEDs are available in several shades that have a color balance typically measured in degrees kelvin (K), falling within the range shown at the bottom of Figure 2-8. Choose a shade that matches your application. Daylight white, at about 4,800 K, appears closest to noonday sun. If your LEDs will be used in combination with "daylight balanced" photography lights, then be sure to use daylight white LEDs. Warm white LEDs, at about 3,000 K, simulate the appearance of older incandescent lamps or early morning sun. Cool white LEDs, at about 7,000 K, appear much bluer and best represent fluorescent lamps or twilight sky.

Color quality is also something to consider. True white light, like sunlight, has roughly equal amounts of all colors from the visible spectrum. However, white LEDs vary in their ability to reproduce the full color spectrum.

Less expensive white LEDs emit spikes of red, green, and blue, without many of the wavelengths in between. Higher-quality LEDs reproduce more colors of the spectrum, thus rendering more colors realistically. As a result, skin tones will appear more lifelike, and teals and purples closer to their actual appearance. There are several methods used to describe LED color quality, including the CRI (Color Rendering Index) and CQS (Color Quality Scale). If you value higher-fidelity color reproduction, and especially if you'll be photographing your lit models, then choose LEDs with a higher color quality rating. Likewise, avoid RGB LEDs that simulate white by combining separate red, green, and blue LEDs in a single package. Another quality factor to look out for is the consistency of hue across the viewing angle. Some white LEDs tend to have yellow or blue halos around the center of their output field.

THE FOUR CS OF DIY LIGHTING

Now that you have a basic understanding of LEDs, here's a simple approach to DIY lighting. Just follow the four Cs:

1. Choose your power supply.

2. Choose your components.

3. Calculate the resistor values.

4. Connect everything together.

Note, however, that LEDs can often be purchased presoldered to wires, sometimes with resistors included. This can allow you to bypass some of these steps, saving you the frustration of soldering the tiny wires yourself. Search sites like eBay for "prewired LED" to see what's available.

Choose Your Power Supply

Start by choosing a power supply that outputs a supply voltage and supply amperage appropriate for your needs. For lighting one or two LEDs, a single 3 V battery will work. But be aware that as batteries age, their supply voltage drops. Therefore, small batteries won't stay bright for long, and you may need to change batteries frequently. A more powerful solution is to use a 4.5 V battery pack. Going to 5 V is an even better choice since you can use a standard 5 V USB battery or wall charger. Higher supply voltages can be used if there's a compelling reason—for example, if your LEGO creation already incorporates the 9 V LEGO Powered Up system to run motors or if you want to use the wide range of prewired 12 V LEDs made for the automotive market. As the supply voltage increases, however, you'll need to add more powerful resistors to the LEDs, which will only serve to convert additional power to heat.

Table 2-1 summarizes the common supply voltage options, typical sources, and advantages to each. Once you've decided which supply voltage is right for your project, search eBay, Amazon, Digi-Key, or a similar retailer to source an appropriate battery holder or wall transformer to supply the power.

Table 2-1: Popular LED Supply Voltages

VOLTAGE	SOURCE	ADVANTAGES
3 V	• 1x CR1225 battery • 2x CR2032 batteries • 2x AA or AAA batteries • Wall power supply	• Standard for consumer electronics • Simplest way to power one or two LEDs
4.5 V	• 3x AA or AAA batteries • 3x LR44 batteries • Wall power supply	• Standard for many toys • LEDs stay brighter for longer as batteries drain and voltage drops
5 V	• USB plug • USB battery • Wall power supply	• Standard for the computer industry • Wide variety of inexpensive USB batteries and accessories available • High enough voltage to support lighting effect controllers and still keep LEDs bright

(continued)

Table 2-1: Popular LED Supply Voltages *(continued)*

VOLTAGE	SOURCE	ADVANTAGES
9 V	• 1x 9 V battery • Wall power supply	• Standard for consumer electronics • High enough voltage to supply many hundreds of LEDs • Same voltage used for some LEGO lighting systems
12 V	• 8x C cell batteries • Automotive plug • Wall power supply	• Standard for the automotive industry • Wide variety of prewired LEDs and accessories (made for the automotive market) available

Choose Your Components

Go to a site like eBay, Amazon, Jameco, or Digi-Key, or just google LEDs, and you'll find hundreds of choices, including the latest and brightest components. Search for "LED" and the type and color you desire. As you review your options, remember the considerations outlined here, including size, brightness, and viewing angle.

For more powerful LEDs, search for "LED" and "1 W" or "3 W" or "5 W." Note that these powerful LEDs often require a heat sink to dissipate the considerable heat they generate, so try to purchase them premounted with a heat sink. Make sure that the seller supplies the specifications for the LEDs you purchase, including input voltage, forward voltage, and current.

Calculate the Resistor Values

To calculate the value of the resistor needed for each LED, use one of the dozens of calculators online that will do this for you. Search for "LED resistor calculator" and just enter the supply voltage, forward voltage, and forward current used by the LED. The calculator will recommend the resistor value and wattage needed. Many LED suppliers also sell the resistors needed for their LEDs and will be happy to consult with you. As noted earlier, prewired LEDs often come with the necessary resistor already soldered in line for you. Just make sure that the rated operating voltage matches your supply voltage.

Connect Everything Together

Now comes the fun part of soldering all your components together. You'll need a soldering iron, electrical solder, flux, wire, and basic soldering skills. Search online for soldering tutorials; you'll find dozens of good videos that teach basic soldering skills. Be careful to use heat sinks where needed, as LEDs are sensitive to overheating.

First, measure and cut the lengths of wire needed to connect the LEDs to the power supply. Be sure to account for twists and turns to route each wire through the MOC. Then solder the wires to the LEDs along with any needed resistors. Carefully work each wire through the MOC. Next, solder the ends of the wires to the battery pack or wall power leads, carefully noting which is positive and negative. When everything's ready, apply power and watch your LEDs light up!

COMMERCIAL LIGHTING

If you don't want to solder your own components together, there are many commercially available LED lighting products that can easily be used with LEGO. Start by looking at your local retail, hobby, and home improvement stores for specialty lighting products. You'll find LEDs are sold to light everything from Christmas trees to kitchen cabinets. Some commonly available commercial products are shown in Figure 2-9.

RETAIL STORE OPTIONS

Miniature hobby lights are short strings of battery-powered LEDs. These are inexpensive and typically feature a tiny LED encapsulated in a plastic bubble every couple of inches on stiff, solid-core wire. Retail stores often sell them during the holidays and place them on clearance afterward. Christmas tree lights are another good option to consider, especially smaller LED string lights with 3 mm or

Figure 2-9: Examples of commercial LED lights: tea light (top left), fluorescent blacklight (top right), multicolor strip light with remote control (center left), COB LED with dimmer (center right), miniature Christmas tree lights (bottom left), and string/fairy lights (bottom right)

5 mm discrete LEDs. These are particularly useful for LEGO models because the lights fit standard-size holes in LEGO elements. Also don't pass up the inexpensive LED tea lights (found in the candle aisle), which can easily be placed in buildings as a quick lighting solution.

STRIP LIGHTS

Strip lights feature LEDs mounted close together on long, flexible, adhesive strips that can be cut to length. They make lighting large surfaces—like the inside of a tall building or the underside of a Technic car—a breeze. Strip lights can typically be found in home improvement stores, where they are sold as under-cabinet lighting; in automotive stores for in-car effects lighting; and at electronics stores, where they provide decorative lighting behind flat-screen televisions. Often, strip lights feature RGB LEDs that can be set to almost any color, usually by a small infrared controller. Some strip lights can cycle through different colors.

ELECTROLUMINESCENT WIRE, SHEETS, AND FILAMENT LEDS

Electroluminescent (EL) wire and sheets (also known as *cold-cathode lamps*) are long wires or sheets that glow when a current is applied. They're available online in several colors. EL wire can be wrapped around bricks to simulate a neon sign or decorate a spaceship (see the *Cyberpocalypse* and *Flying Delorean* models on pages 80 and 117, respectively).

EL sheets can backlight a window or rocket engine. However, EL components are relatively dim and so work best in very dark locations.

All EL components are driven by an *inverter*, an electronic component that takes an input voltage, increases it, and pulses it. Sellers of EL wire usually sell matching inverters, too.

In many of the scenarios where you'd want to use EL wire, you could use filament LEDs as an alternative (see the section "Brickstuff" later in the chapter). Filament LEDs are a newer, very bright option in which many very tiny LEDs are embedded in a thin, flexible wire with a rubberized coating. Filament LEDs are much brighter than EL wire, don't require an inverter, and can be bent around tight corners. They do, however, draw a lot of current.

OPTICAL FIBER AND PANELS

Plastic sideglow optical fiber, lit by LEDs on the ends, is an alternative when EL wire isn't bright enough. However, while optical fiber can be brighter, its brightness diminishes along its length, whereas EL wire's dimmer glow is more consistent across its length. Plastic sideglow optical fiber can purchased online in various thicknesses.

Similarly, plastic *edgelit* or *sidelit* LED panels glow when lit by LEDs from the side. These range from small, inexpensive panels used to backlight cell phones to large commercial ceiling panels that can light an entire room. The LED light panels that artists use to trace and copy images are an inexpensive option that can be found in art stores; they typically feature a sheet of transparent plastic that is edgelit by one or more strips of white LEDs. Similar products are available in photography stores for viewing slide transparencies.

LED light panels produce an even, diffuse light that is perfect for certain situations. For example, placed high above a build, they can simulate ambient skylight. They can also be placed behind translucent bricks to back-light a stained glass window. They can even be laid flat and covered with trans-orange plates to mimic the appearance of lava.

EDGE LIGHTING

Edge lighting involves embedding tiny lines or holes in a clear plastic panel to reflect light 90 degrees from the side of the panel to its face. This kind of lighting is widely used for laptop, tablet, and smartphone displays. Such displays can be repurposed for LEGO lighting, but usually at a higher cost and effort than artists' light panels.

THIRD-PARTY LIGHTING SYSTEMS

Lighting LEGO creations has become so popular that a small industry of third-party lighting companies has emerged to sell custom lighting solutions well adapted to the needs of LEGO enthusiasts. These companies offer integrated systems of plug-and-play components that let you quickly and easily install bright, colorful, and reliable lights into your LEGO builds. Their components are often available individually so that you can purchase only what you need to light your MOCs, or they come in lighting kits designed specifically to light your favorite LEGO sets. The latter include everything needed for installation: prewired LEDs, connecting wires, and instructions. Just add a USB power supply or battery, and you're in business. The premium companies even include LEDs preinstalled in genuine LEGO bricks. Value-based companies often use clone bricks. The more innovative companies also offer devices for controlling sound and animated lighting effects, which will be discussed in detail in Chapter 9.

Many companies sell lighting systems compatible with LEGO. This section will discuss five of them. The first three—LifeLites, Brick-stuff, and Brick Loot—are big innovators in the LEGO lighting field. Over the last several years, a lot of clone LEGO lighting companies have popped up. Many have imitated these three companies' products, but none can lay claim to the innovation and originality that these manufacturers have contributed to the LEGO fan community. Light My Bricks, a newer manufacturer, is also profiled here due to its large product line and market presence. Last is the Woodland Scenics Just Plug system. This system was developed for the model railroad industry, but it works very well with LEGO, too.

LIFELITES

In 2004, LifeLites of Bismarck, Arkansas, became the first third-party company to offer lighting components specifically designed to work with LEGO bricks. Rob Hendrix, the founder and an avid LEGO enthusiast, has long made custom lighting for other AFOLs. His passion and experience led to several other firsts: the first battery to fit into a 2×4 brick, the first 2×2 battery brick, and the first lighting effect controller (LEC) designed specifically for LEGO. In fact, as of this writing, Life-Lites' eLite Advanced LEC, which is discussed more in Chapter 9, remains the most powerful effect controller on the market for its size.

Figure 2-10 highlights some LifeLites products. The 2×4 brick (top right) is available

as a distribution strip or LEC version. The cables are robust and feature large, easy-to-handle plugs.

The LifeLites eLite Series is ideal for vehicle lighting. Primarily designed for battery power, this 3.4 V system has compact components that fit into the space of a 2×4 brick. The system offers 3 mm through-hole LEDs in a range of colors that fit perfectly into LEGO headlight bricks. The ModuLite series for interior lighting features SMD LEDs. The NanoLite line features LEDs so small that they fit inside the hole on the underside of a 1×1 round plate and thin flex ribbons that run between plates. Best of all, NanoLites feature the same micro connectors used by Brickstuff and Light My Bricks, so these products can be used together.

BRICKSTUFF

In 2011, Rob Klingberg, another AFOL who loves to tinker with lighting, founded Brickstuff in Minneapolis, Minnesota. A company dedicated to premium, innovative LEGO lighting solutions, Brickstuff was the first to offer bright strip lights for building interiors, tiny Pico LEDs small enough to fit inside 1×1 plates, and wires thin enough to fit between bricks. Best of all, Brickstuff's tiny connectors fit through 3 mm holes, allowing the wires to be strung between all manner of bricks and plates.

A selection of Brickstuff's premium DIY components is shown in Figure 2-11. The standard Pico LED (bottom left) is small enough to fit

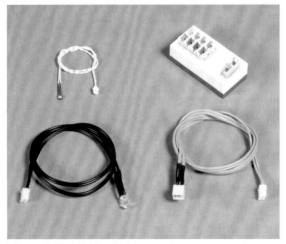

Figure 2-10: LifeLites lighting system components

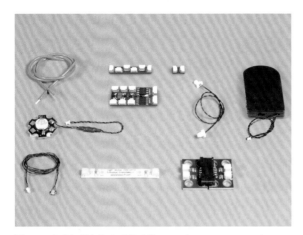

Figure 2-11: Brickstuff lighting system components

under plates. The High-Power LED (middle left) contains a super-bright 1 watt LED. The Light Strip (bottom center left) with sticky tape on the back is an ideal solution for lighting interior rooms. The Filament LED (top left) is a bright, flexible, glowing wire. Brickstuff standardized the use of very thin multistrand wire (see the Pico LED in the lower left and the extension wire to the left of the black battery box).

Over the years, Brickstuff has developed an innovative array of compatible accessories, including LECs, sound boards, video displays, and LEGO-themed EL backlit signs. It also offers a high-power analog dimmer that can provide flicker-free dimming for hundreds of LEDs. The company's impact on the industry can be seen in the products of newer lighting suppliers who have adopted Brickstuff's components and standards. For example, other manufacturers make lights that match the Brickstuff Pico LEDs but are known by other names. This book will use the term *pico* generically to refer to any manufacturer's light that fits inside a 1×1 plate like the Brickstuff Pico LED.

BRICK LOOT

In 2014, the Brick Loot Company of Chicago became the exclusive North American distributer for LiteUp Blocks, the first company to manufacture value-based lighting kits custom-made for specific LEGO sets. Parker Krex, Brick Loot's CEO, remarkably founded the business with his family when he was just nine years old. Brick Loot's philosophy is to make lighting easy for the customer through components like the ones shown in Figure 2-12. Brick Loot's lights are bright, many with multiple LEDs on a single board mounted inside a 1× plate.

Brick Loot's Light Linx system, introduced in 2019, allows you to connect (daisy-chain) all lights in series instead of having to plug each individual light into a USB hub. This can greatly reduce the number of wires that you have to run for large installations. It also facilitates multi-circuit design (see "Multi-Circuit Design" in Chapter 4).

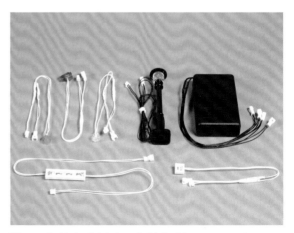

Figure 2-12: In Brick Loot's Light Linx system, the lights (top and bottom rows) are designed to be daisy-chained to save space. The battery case (upper right) has multiple plugs to aid with multi-circuit design.

Brick Loot LEDs are 5 V surface-mounted to small boards with solid-core copper wires. These solid-core wires are less expensive than multistrand wires, but be aware that crimping them tightly several times in the same place can eventually cause them to break. This is not a problem if you plan to keep the lighting kit installed permanently. But if you want to reuse these components in other builds, be careful bending the wires—and expect some to break.

LIGHT MY BRICKS

Founded in 2016 in Melbourne, Australia, Light My Bricks (LMB) adopted the same system standards originated by Brickstuff, and many of the components are compatible. The LMB portfolio covers a wide array of LEGO sets. LMB also has a reseller program, so their sets and components sell through many other retailers around the world. LMB offers LEDs called Bit Lights, cables, a pulse-width modulation (PWM) dimmer, and distribution boards compatible with Brickstuff. The company also offers Micro Bit Lights, which feature a tiny LED smaller than 1 mm, and Large Bit Lights, which offer a larger, brighter LED. These and other components are shown in Figure 2-13.

WOODLAND SCENICS

In 2016, longtime model railroad manufacturer Woodland Scenics introduced the Just Plug LED lighting system, shown in Figure 2-14. This system is unique among those listed here in that it is not specifically intended to light LEGO creations. It is, however, well designed

for lighting miniature towns and perfectly usable for LEGO, too. It has quality components designed to just plug together.

Just Plug Stick-On Lights use a chip with three individual LEDs that are very bright and can light an entire hollow building. The Just Plug line also offers tiny nano lights and spotlights. But what makes the Just Plug system

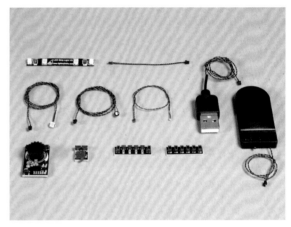

Figure 2-13: Light My Bricks lighting system components. The Strip Light (top row, left) comes in several colors. The Bit Lights are similar to Brickstuff Pico LEDs and come in Large (middle row, second from left) and Micro (middle row, third from left).

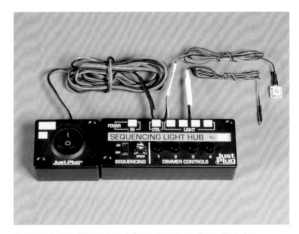

Figure 2-14: Woodland Scenics Just Plug lighting system components. The on-off switch (left) can be positioned anywhere, and the lights (top right) come in large and small versions.

special is that it provides analog dimming for every LED as standard, allowing multiple strings of LEDs to be economically dimmed and balanced. This is especially useful for Brickfilms makers who need to adjust lighting.

COMPARING THIRD-PARTY LIGHTING KITS

Several third-party lighting companies sell lighting kits made for specific LEGO sets. They contain all the necessary LEDs, wires that are precut to length, plugs to connect everything together, and instructions that show you how to install everything. Such kits are a quick, economical way to light your favorite LEGO set, but how do you choose which manufacturer's lighting kit to purchase? Start by comparing each kit's design, LED color purity, quality of components, special features, and cost.

Lighting Kit Design

Lighting kits vary widely in the number and placement of LEDs and in the way the LEDs are integrated. Some kits use bright, saturated colors, while others use a more prototypically accurate lighting style. Some kits provide just basic ceiling lights, whereas others incorporate unique lighting features like candles, fireplaces, lanterns, and spotlights.

Looking at photographs of an installed lighting kit is a great way to assess a manufacturer's design. Be careful, however, when looking at photos supplied by manufacturers, as they employ professional photographers. Make

sure that you're seeing the LEDs themselves and not just great studio lighting. Look for unedited videos and photos online, or see if you can view these lighting kits in person at LEGO fan shows near you, before making a purchase decision.

Figure 2-15 shows lighting kits from three different manufacturers for the popular LEGO set #71040 "The Disney Castle." These photographs were all shot under identical lighting conditions and camera settings. Consider the brightness, number, and placement of LEDs. Note that these photos don't convey the animated lights, which cycle through various colors. Also note that interior lights will unavoidably "bleed" through thin elements (like the #87421 3×3×6 Corner Wall Panel) due to the opacity of the plastic used by LEGO. Considering all these factors, lighting design is very subjective. Which looks best to you?

Quality of Components

Consider the durability of the LEDs, wires, and connectors when choosing a third-party kit. Components from premium manufacturers are designed to hold up when reinstalled many times. The solid-core wire and cheaper connectors found in value-based manufacturers' products will work fine for a single installation but may not handle repeated reinstallations into other builds. If you envision installing a lighting kit in your favorite LEGO set and placing it on the shelf forever, then value-based lighting is fine. But if you envision reusing the lighting components in various MOCs, then spend the extra money and get premium components.

Figure 2-15: Comparison of lighting kits for LEGO set #71040 "The Disney Castle." The top-left photo shows the castle with only dim ambient room lighting. The top right shows the Brick Loot lighting set; the bottom left, the Light My Bricks lighting set; and the bottom right, the Brickstuff lighting set.

Another quality factor to consider is that premium manufacturers' lighting kits typically feature LEDs installed in genuine LEGO brand elements. Lighting kits from value-based manufacturers often use clone bricks. In many cases the difference is small, but it may matter to a LEGO collector.

Special Features

Premium lighting manufacturers are now bundling special features into their lighting kits. These include remote control, sophisticated lighting animation, and sound. These extra features are difficult to compare, but keep them in mind when choosing kits. The manufacturers often have videos that show animation on their websites.

Cost

Last but not least, consider the cost of each manufacturer's components. Premium-priced lighting components cost about the same as current LEGO Power Functions or Powered Up lights (about $6 per LED as of this writing). Note that Woodland Scenics products are widely discounted through hobby shops and online sellers. Value-priced lighting kits cost slightly less than buying LEGO 4.5 V and 9 V lights on the open market (about $1.50 per LED as of this writing). Of course, you have to add in the cost of wires, distribution hubs, power supply, any LECs, and your time. Therefore, a typical small building installation (eight LEDs with cables, boards, and battery pack) will cost about the same using any of the premium manufacturers' components.

There are three hidden Easter eggs in the photograph at the beginning of this chapter: (1) The meter in the lower left reads 0637, which is *LEGO* upside down. (2) The soldering station on the top left displays 245, the number of the first LEGO lighting set. (3) The calculator on the right displays 40237, the LEGO set number for . . . an actual Easter egg!

PROJECT 2:
BUILD YOUR OWN LEDS

Have you ever wanted to build oversized electronic components out of LEGO? Now you can, while learning how LEDs work at the same time. This project contains instructions to build two LEDs wired up on an oversized breadboard (see Figure 2-16). Add even more oversized LEDs in different colors and impress your friends with your understanding of basic electronics.

Requirements

- Two pico LEDs that can fit under a 1×1 round plate and have wires thin enough to run between bricks, available from Brickstuff or an equivalent third-party lighting manufacturer

- One adapter board with at least two connectors, as required to connect the LEDs to the battery pack

- One battery pack small enough to fit into the single-stud height underneath the model

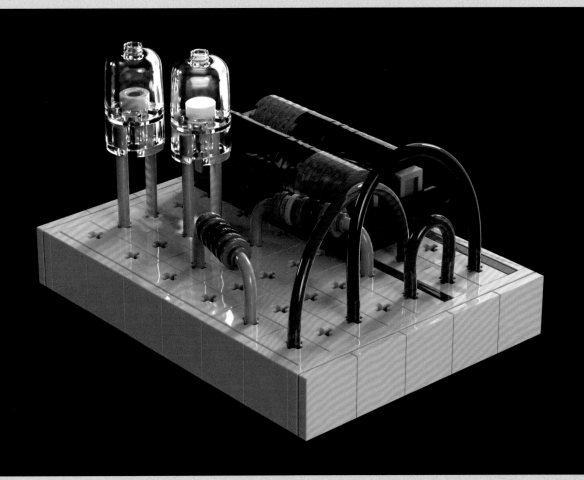

Figure 2-16: Project 2: Custom LEDs

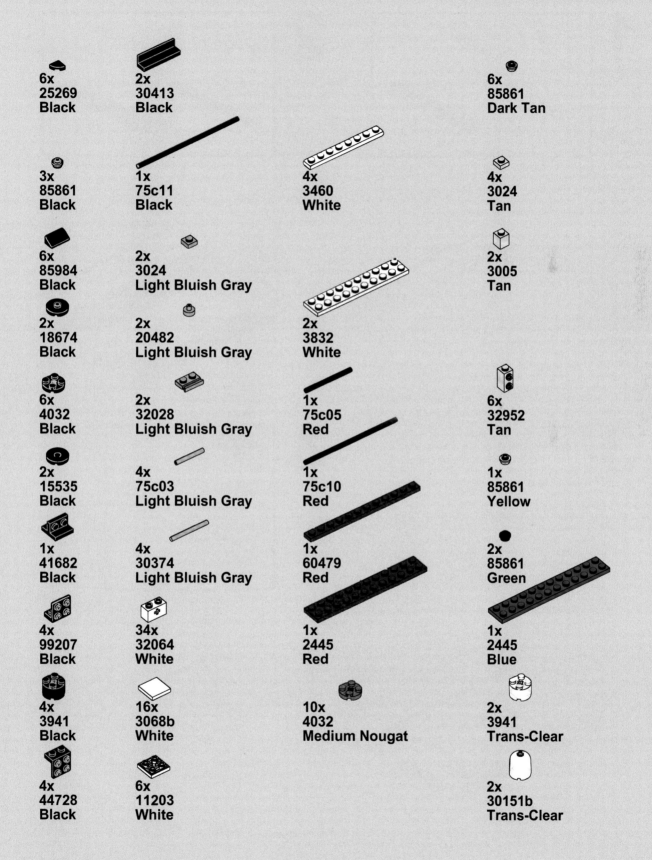

6x
25269
Black

2x
30413
Black

6x
85861
Dark Tan

3x
85861
Black

1x
75c11
Black

4x
3460
White

4x
3024
Tan

6x
85984
Black

2x
3024
Light Bluish Gray

2x
3005
Tan

2x
18674
Black

2x
20482
Light Bluish Gray

2x
3832
White

6x
4032
Black

2x
32028
Light Bluish Gray

1x
75c05
Red

6x
32952
Tan

2x
15535
Black

4x
75c03
Light Bluish Gray

1x
75c10
Red

1x
85861
Yellow

1x
41682
Black

4x
30374
Light Bluish Gray

1x
60479
Red

2x
85861
Green

4x
99207
Black

34x
32064
White

1x
2445
Red

1x
2445
Blue

4x
3941
Black

16x
3068b
White

10x
4032
Medium Nougat

2x
3941
Trans-Clear

4x
44728
Black

6x
11203
White

2x
30151b
Trans-Clear

1

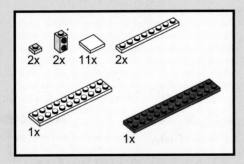

2x 2x 11x 2x

1x 1x

2

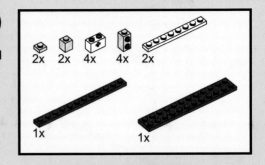

2x 2x 4x 4x 2x

1x 1x

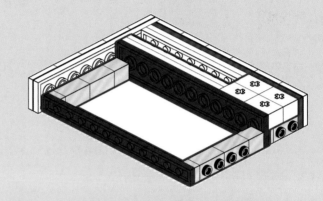

3

30x

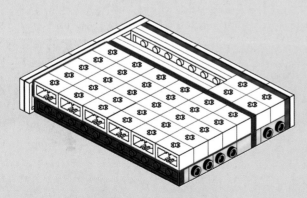

4

6x 5x

1x

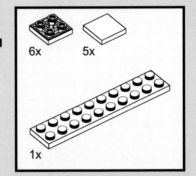

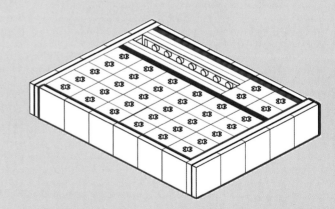

5

4x   4x

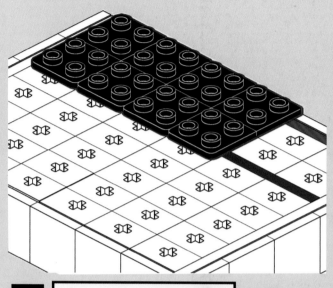

6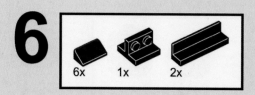

6x   1x   2x

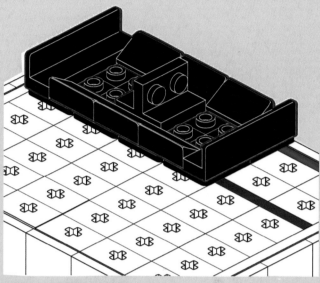

7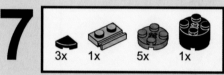

3x   1x   5x   1x

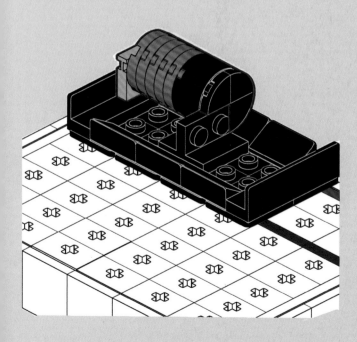

8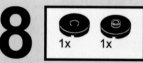

1x   1x

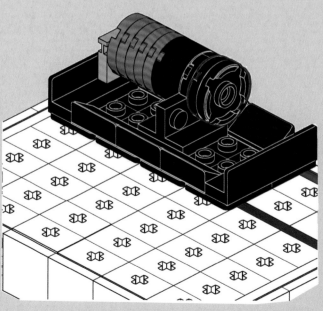

9

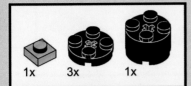

1x   3x   1x

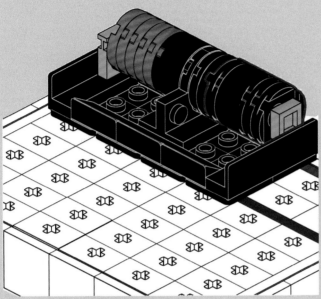

10

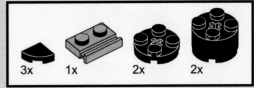

3x   1x   2x   2x

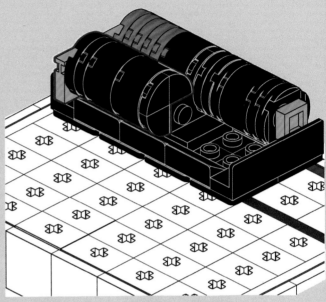

11

1x   1x

12

1x   1x   5x

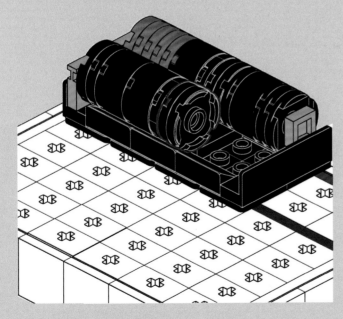

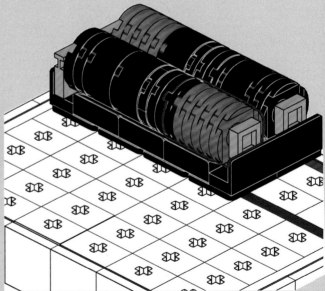

13

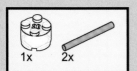

1x   2x

14

1x   1x

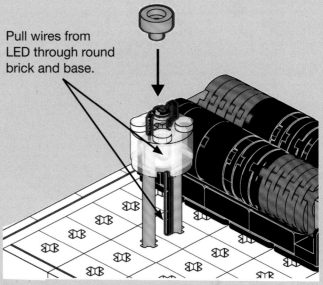

Pull wires from LED through round brick and base.

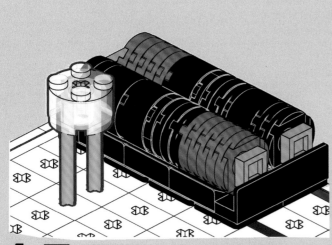

15

1x

16

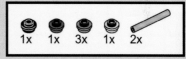

1x   1x   3x   1x   2x

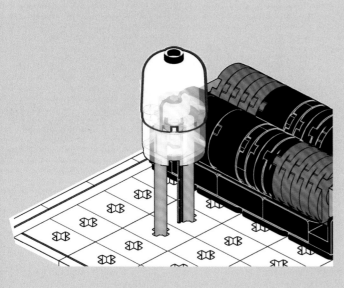

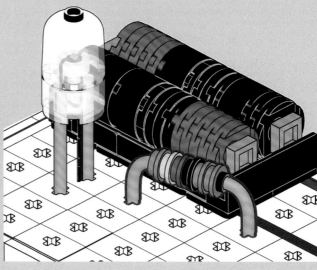

17

1x   2x

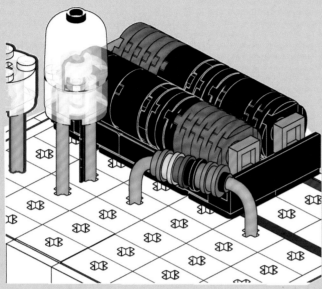

18

1x   1x

Pull wires from
LED through round
brick and base.

19

1x

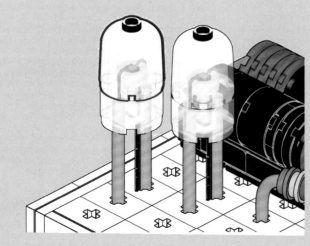

20

2x   3x   1x   2x

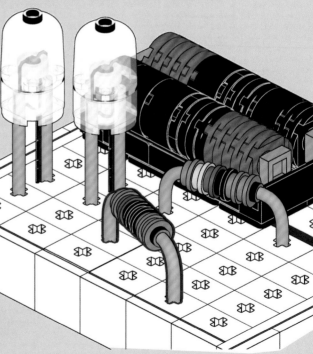

21

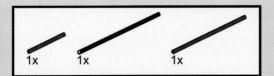

1x 1x 1x

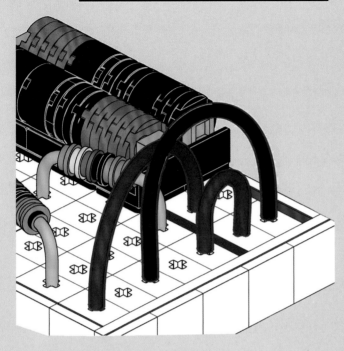

22

1x

Gather wires.

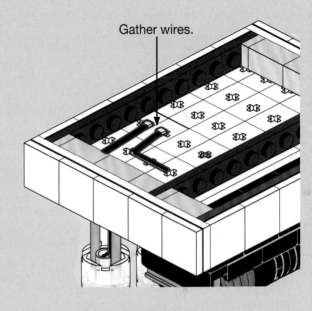

23

1x

Connect wires to board. Connect board to power supply.

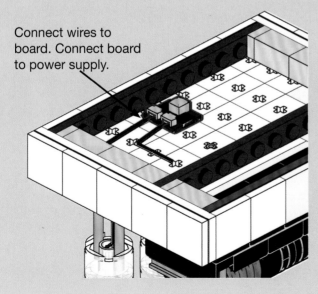

24

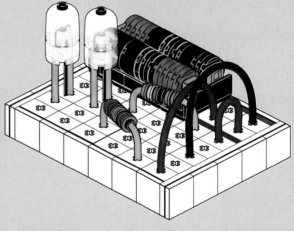

Library of Druidham (Benjamin Stenlund)

3 INTERIOR LIGHTING

Lighting is everything. It creates atmosphere, drama, and intrigue in a room.

—MARTYN LAWRENCE BULLARD,
INTERNATIONALLY ACCLAIMED
INTERIOR DESIGNER

Lighting adds timeless appeal to miniature rooms. Incandescent lights first appeared in dollhouses early in the 20th century. Some of those elaborate miniatures rise to the level of art and survive to this day, such as actress Colleen Moore's *Fairy Castle*, on display at the Chicago Museum of Science and Industry, where it continues to enchant new generations. With the rise of the motion picture industry came Hollywood visual effects pioneers like Ray Harryhausen, who built, lit, and even destroyed miniature movie sets, all for the camera. Hollywood miniature lighting and photography were later refined to delight audiences of some of the most popular motion picture franchises, like *Star Wars*, *Harry Potter*, and *The Lord of the Rings*.

Today, 3D computer software has largely replaced the use of miniature effects. But the virtual lighting used for digital models is based on the same principles used to light real-world sets. Those principles are just as applicable in lighting miniatures made from LEGO. In fact, one of the most popular uses for LEGO lighting is to illuminate interior scenes, typically the minifig-scale rooms contained in premium LEGO sets. You'll find the widest range of creative possibilities, however, in lighting the custom MOCs that you build yourself. You can add as few or as many lights as your imagination can illuminate.

THE ART OF CINEMATIC LIGHTING

The simplest way to start lighting an interior scene is with a single light on the ceiling in the center of the room. This *basic lighting* illuminates the room evenly, allowing the viewer to see everything. Through more careful control of the light, however, you can add greater interest to the scene, tell a story, and achieve *cinematic lighting*. Imagine yourself as a Hollywood director of photography who is lighting a movie set. Use light creatively to build the mood of the scene and draw the viewer's eye to the main actions. Consider the contrast between basic lighting and cinematic lighting shown in Figure 3-1.

Figure 3-1: *Night Before Christmas* (author) with basic lighting (top) and cinematic lighting (bottom)

The scene with basic lighting uses a single LED to illuminate the room evenly. It allows the viewer to see all the details but misses several opportunities to add character and charm to a Christmas scene like this.

Compared to the basic lighting setup, the cinematic lighting version more effectively conveys mood and directs the viewer's attention, helping tell a story. The scene is darker, allowing the warm glow of the fireplace to become the primary source of illumination. The backdrop outside the window is now lit, revealing the falling snow. Tiny accent lights highlight the Christmas tree and chair. Finally, lights draw the viewer's eye to Santa, who is apparently trying to quietly appease the family dog who has greeted him. See how much more dramatic and interesting this interior scene becomes with cinematic lighting?

SHADOW BOX DIORAMAS

The *Night Before Christmas* scene shown in Figure 3-1 is an example of a *shadow box diorama*, a type of miniature scene that enables great storytelling by giving you maximum control over viewing angle and lighting. A shadow box diorama depicts a scene, often an interior room, housed in a box that conceals lights, wires, and other elements and permits a single viewing angle through the front (see Figure 3-2). The scene is usually set back in the housing to provide a darkened interior where the lights are easy to see.

Many LEGO sets feature buildings with interiors that are open for viewing on the backside, like a shadow box diorama. LEGO sets with floors that stack on one another (that is, multistory builds), like those in the Café Corner series, are better suited for viewing through their windows from the outside, however, as the backsides are not open. But you can rebuild them that way if you like.

THE ART OF SHADOW BOXES

Prime examples of the artistry of traditional shadow box dioramas are the Thorne Rooms at the Art Institute of Chicago and the creations of renowned master military diorama builder Sheperd Paine.

Figure 3-2: *Night Before Christmas*, showing shadow box diorama details

INTERIOR LIGHTING STYLES

Lighting rooms made from LEGO bricks is a lot like lighting full-size rooms. Therefore, you can draw upon the four traditional lighting styles employed by interior decorators for ages: natural, ambient, accent, and task lighting.

NATURAL LIGHTING

The first method of lighting a LEGO room is to simulate natural light coming in through windows. Skylights can create illumination from above (as in the scene shown in Figure 3-3), and a single wall of windows can bathe an entire room in warm sunshine or backlight a scene for effect. The location and size of the windows controls the mood, along with the type of light you use. The light can range from soft blue moonlight to yellow sunlight at dawn.

Bright, hard-edged light can create the awe-inspiring shafts of light pouring through windows that acclaimed movie director Steven Spielberg calls "God Lights." Natural light coming through windows is easy to implement in a shadow box diorama because the outer box hides the lights and backdrop. For example, the blue snow backdrop used in Figures 3-1 and 3-2 and the lights illuminating it are hidden between the inside and outside walls of the scene.

AMBIENT LIGHTING

Ambient lighting provides general soft illumination to fill a room without shadows. Often

Figure 3-3: *Hydroponics Research & Development Facility* (Jon Blackford) uses natural light from skylights to create ambient lighting.

this comes from the cumulative effect of several lights. You can recess strip or down lights in the ceiling and even angle them to direct light backward. You can use floor lamps and wall sconces to bounce light off the ceiling or hide lights behind chairs and other objects. If there's enough natural light, it can reflect off the floor and walls to fill a room with ambient light, too, as with the skylights in the hydroponics facility in Figure 3-3.

ACCENT LIGHTING

Accent lights draw attention to specific features or objects in a room, such as spotlights focused on artwork, plants, or bookcases or wall sconces placed artistically on wall panels. Decorative lights inside display cases or strip lights used above cabinets to define ledges also fall into this category.

TASK LIGHTING

Task lighting uses light to illuminate areas where work is typically done, like a ceiling light over a desk, under-cabinet lights over a kitchen counter, or a reading light next to a chair. The example in Figure 3-4 uses lights hidden in the ceiling to direct pools of light onto the incubators and other work areas.

Figure 3-4: *Bio Lab One* (Wami Delthorn) shows task lighting.

The lighting in the freezer case in the back of the scene is another form of task lighting. Notice also the soft, even illumination from the front—this is an example of how task lighting is often combined with ambient light to provide general illumination.

CINEMATIC LIGHTING TECHNIQUES

Interior lighting illuminates a room, but cinematic lighting illuminates the heart. Lighting rises to a whole new level when it helps tell a story. When creating cinematic lighting, first determine the story that your scene is meant to tell. Once again, try to think like a Hollywood director of photography. If the scene you are lighting is from a movie, video game, or photograph, study the original closely. What is the scene's focus, and how does the lighting support that focus? Where are the sources of light? Are they offscreen, or are there visible light sources such as lamps or candles?

The wide variety of available LEGO minifigs makes LEGO bricks a great medium for telling stories. Even the best-built LEGO scenes with lots of detail and lighting remain static until the story within is realized. A bank, for example, can become the setting for a robbery. A living room can become the setting for a birthday party, and an ordinary aircraft hangar can become the setting for a secret government conspiracy to fake a moon landing. Be creative. Find the story you want to tell and consider how lighting can help you tell it.

First, identify the main action of your scene, which is typically one or more minifigs doing something. It could be police officers arriving at the bank, flashlights in hand, to discover that all the vault's money, like the robbers, is long gone. Or it could be a little girl blowing out the candles on her birthday cake, or a famous Hollywood science fiction director having his crew place lights to make a soundstage look just like the moon.

Next, use the *three-point lighting technique* widely employed in photography and cinema to highlight the subject using three light sources: a hard-shadow *key light*, the main source of illumination, placed on one side of the camera (or MOC viewer, in this case); a diffuse *fill light*, placed on the opposite side; and a *backlight*, or *hair light*, placed in back above to add highlights to the head and shoulders (Figure 3-5). To add drama, move the backlight to one side and lower it below the shoulders, where it would be more properly called a *kicker light*.

Note that hiding the three light sources in a scene can be a challenge, especially for

Figure 3-5: The three-point lighting technique

of a room with reds and oranges heightens the tension and communicates danger, action, or passion. Projecting blue can communicate calm or hope. Green can be eerie and foreboding. If you need to adjust the color of lights finely—for example, to get the right steel-blue hue of moonlight—consider filtering white light with trans-colored plates or gels (see the box "Light Modifiers" later in the chapter). Another alternative is to use RGB LEDs that can project a wide variety of colors.

rooms with low ceilings. In this case, consider building your floor extra thick to hide the lights within. Another good technique is to hide lights behind furniture or even minifigures.

Cinematic lighting also harnesses color to create a mood. As a builder, you can choose to use colored bricks or vary the color in a scene using lights. Lighting the background

LIGHT FIXTURES

There are many ways to combine LEDs with LEGO bricks to create functional miniature light fixtures to illuminate your scenes. Key and fill lights are typically hidden above a scene, or located "offstage" so to speak. *Practical lights* are any lights visible in a scene, including everything from fireplaces to floor lamps. This section describes LEGO light fixtures you can build to implement interior lighting styles and cinematic lighting techniques.

KEY LIGHTS

A key light, as mentioned earlier, provides the primary illumination for a subject and is usually the brightest in a scene. The light fixtures shown in Figure 3-6 are useful designs for key lights.

Figure 3-6: Key lights

Install a pico LED with a wide viewing angle in the fixture by placing a 1×1 round brick over it to restrict the projected light. Or you can incorporate a 9 V light reflector dish (BrickLink #x85) to reflect most of the light forward. You can make a large, powerful key light from a 1 watt LED.

HANGING BRICK LIGHTS

LEGO lights should be mounted on hinged joints so you can position them with precision. You can hang them from a ceiling, mount them onto columns, or attach them to the floor. If you need to use several lights, consider building an overhead grid from Technic beams or axles. The grid will give you lots of flexibility to reposition lights and wrangle wires.

Discrete LEDs can work for small key lights because the LED package is designed to focus most of its light forward within a 30-degree viewing angle. Headlight bricks make good housings because they provide a tight fit for 3 mm LEDs and have a second stud to attach *barn doors*, flaps that restrict the light to the right area.

FILL LIGHTS

Fill lights, such as those shown in Figure 3-7, are effective for general illumination and filling in shadows.

Fill lights produce soft shadows, which you can often create by turning an LED backward and bouncing the light off a white element (such as a 2×3 dish), mimicking how umbrellas diffuse light in real studio lights. You can create small fill lights from small LEGO windows by bouncing the light off the white-colored glass element in a window frame turned backward. For large fill lights, use 1×6×5 LEGO panel elements. To better diffuse the light, it helps to cover

Figure 3-7: Fill lights

the top of a fill light with a diffusion-type photography lighting gel, which is a thin, semi-opaque sheet of plastic that scatters the light that passes through it. If you don't have a diffusion gel, try using Scotch Magic Invisible tape.

LIGHT MODIFIERS

Light modifiers are materials and devices used to alter light's appearance. A white panel used to bounce light is one example. You can also cover a panel with crumpled aluminum foil for a different look. Gels are clear plastic sheets that come in a range of colors and are used to change the color of light. White diffusion gels are also available to soften shadows. Look for these at a well-stocked photography store. Consider buying a swatch book, which is a pack that contains small samples of each color gel the manufacturer offers. Swatch books are inexpensive and work well for miniature lights.

If you're making a shadow box depicting an outdoor setting, make an overhead fill light to mimic daylight by combining several down lights or strip lights into a large panel placed over the scene.

CEILING FIXTURES

The easiest ceiling fixture is a simple strip or down light. But you could also build some creative hanging ceiling lamps with just a few LEGO elements, as shown in Figure 3-8.

Just about any LEGO elements can be used for these hanging lights, but the best ones will be trans-colors and refract light like real crystal glass. Warm white pico LEDs usually work best with interior scenes evoking a cordial feeling.

You can light utility and industrial lamps with discrete LEDs or SMD LEDs, depending on the look and amount of light you want. Discrete LEDs create pools of light on the floor due to their narrow viewing angle. SMDs cast light over a wider area and are available in both pico and larger sizes. You can easily hide thin wires without connectors inside a 3 mm hose used to suspend a light fixture. If you're using wires with connectors already attached, then you can slit the 3 mm hose down the side using a hobby knife and slip it around the wire.

Figure 3-8: Ceiling fixtures

Hold LEGO bricks up to a light, and you may notice that some bricks appear more opaque than others. This is true for bricks in most colors, especially lighter colors. Some appear solid and block light coming through, while others appear almost translucent. You can take advantage of the translucent ones for lighting. For example, placing an LED inside a "translucent" white head brick will cause the entire head to glow (see the airplane lamp in the top row of Figure 3-8).

TABLE LAMPS

Table lamps are perfect to add accent and/or task lighting to draw attention to a work area, table, credenza, or chair. Each is powered by a small pico light with carefully hidden wires (see Figure 3-9). You can place the LED facing up or down depending on the type of lampshade and pattern of light you wish to cast.

Figure 3-9: Table lamps

Usually only one or two lamps per room are needed, so don't overload the space, unless you're modeling a familiar scene that requires many lights. Also note that indoor lamps are typically incandescent and therefore should use a warm white LED.

FLOOR LAMPS

Floor lamps are used for general or task lighting (see Figure 3-10). They can provide direct light or bounce light off a ceiling. In the latter case, make sure that your ceiling is made from white or light-colored bricks.

Figure 3-10: Floor lamps

Each floor lamp is powered by one or several pico lights with carefully hidden wires. Like other indoor lamps, floor lamps are typically incandescent and therefore should use a warm white LED. The exception would be lamps modeled after prototypes that use very bright halogen bulbs. These will look more realistic with a cool white LED.

WALL SCONCES

Wall sconces add a decorative flair to walls and provide ambient light in an interior scene (see Figure 3-11). Often, as with real sconces, the light bounces off the ceiling to achieve a diffuse effect.

Wall-mounted lamps can present a challenge for hiding wires if the other side of the wall is an adjacent room or the exterior wall of a building. In this case, try to make the wall two studs thick to hide the backside of the lamp. You can use a 1×2 brick modified with groove (#4216) to run thick wires within the wall. If your wall is one stud thick, you can hide the wire between bricks or behind a bookcase or painting.

Figure 3-11: Wall sconces

ARCHITECTURAL LIGHTING

Architects often incorporate custom lights to reinforce a building's style. For example, Frank Lloyd Wright was fond of wall sconces and designed several in his signature prairie style. Look for unique lighting in prototype buildings and incorporate it into your LEGO creations.

Figure 3-12: Specialty lamps

SPECIALTY LAMPS

Figure 3-12 shows several examples of specialty-use lamps, including a cast-iron stove, a fish tank, and a television. These are all practical lights appearing in scenes not to provide illumination but to provide character that can add interest. All take advantage of trans-colored elements lit with pico LEDs. The lava lamp (top row, second from right) takes advantage of new LEGO glitter trans-colors.

Note that some specialty lamps can look even better when driven with a lighting effect controller, or LEC (see Chapter 9). For example, the stove shown here—like fireplaces, torches, and televisions—could benefit from a slight flicker.

DIMMING LIGHTS

By default, lights are connected directly to a power source and run at full brightness. However, at times you'll want to dim some or all of your lights. Many of the best examples of brick lighting come from the makers of *Brickfilms*: stop-motion animation movies made using LEGO bricks. LEGO filmmakers, like their Hollywood counterparts, often dim lights for both technical and artistic reasons.

Dimming often involves driving LEDs using *pulse-width modulation*—delivering power in a series of pulses rather than continuously—through an LEC. LifeLites, Brickstuff, and Light My Bricks all offer LECs that provide dimming. If you want more choices, a wide range of dimmers can be sourced online, including the Trinket series through Adafruit. LEGO purists can use the old LEGO 12 V and 9 V train controllers that regulate voltage to dim lights. And if you're really serious about dimming, try the Woodland Scenics Just Plug system (described in Chapter 2), which incorporates four separate dimming channels into each light hub.

EXAMPLES OF CREATIVE INTERIOR LIGHTING

Now that you've seen some interior lighting styles, cinematic lighting techniques, and lighting fixtures, let's take a look at how you can combine them to make some amazing MOCs. Study the way light plays a key role in each of these scenes. With a little practice and a lot of patience, you'll soon be creating scenes like these yourself.

FAKE MOON LANDING

The television studio shown in Figures 3-13 and 3-14 features three cool-white spotlights hanging on a lighting rack high above the action. They simulate the cold harsh light on the moon's surface. Several practical lights also help sell the action, including the studio lights, the "On Air" light, and the live camera light. Note that the practical lights are warm white to draw the viewer's eye, and the one on the left is effectively backlit.

Figure 3-13: *Fake Moon Landing* (author)

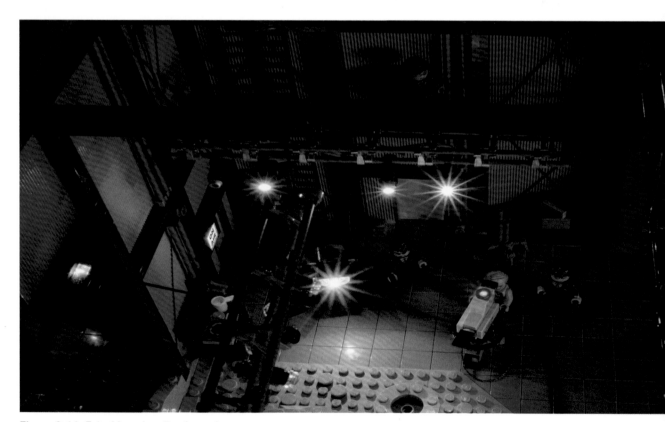

Figure 3-14: *Fake Moon Landing* from above

COFFEE SHOP

Figure 3-15 is an excellent example of how you can use light to craft the morning bustle of a coffee shop. Warm white LEDs generate the sunlight streaming through the windows, which gives the scene its golden glow. The ceiling lamps use pico lights to create ambient light that casts interesting shadows on the wall. Don't overlook the pico lights installed in the pastry and beverage cases to provide task lighting.

Figure 3-15: *Coffee Shop* (Foolish Bricks)

AREA 51 WAREHOUSE

Large warehouses typically feature industrial overhead lamps. The lamps in the scene shown in Figure 3-16 are lit with 5 mm through-hole LEDs scavenged from Christmas tree lights. Note the pools of light these cast on the floor and crates, illuminating the extensive use of custom stickers. Less obvious are the mirrors used to make this scene appear much larger than it really is.

Figure 3-16: *Area 51 Warehouse* (author)

Figure 3-17: *Speeder Repairs* (Benjamin Stenlund)

SPEEDER REPAIRS

The futuristic repair shop in Figure 3-17 has banks of large overhead lamps that provide broad ambient light. This both highlights the extensive details and reveals the different shades of gray used in the walls. A couple of orange practical lights represent machinery. The use of red and purple complementary colors in the rear chamber underscores the mood and draws the viewer deeper into the scene.

FINDING THE ARK

A big challenge of lighting MOCs is concealing the lights. In the scene in Figure 3-18, the highlights and kicker lights are placed far above and off to the sides. The walls are lit from below using a one-stud-wide trough at the base. Finally, and most prominently, a strip light with a yellow gel is positioned in the base of the container holding the ark.

INTERROGATION TENT

The scene in Figure 3-19 shows the difference that a few well-placed practical lights can make in a confined space. All LEDs used in this scene are pico LEDs. The warm white lantern on the right creates an eerie contrast with the cool white lamp on the left. A warm highlight above defines the characters' faces and Indiana Jones's iconic fedora. Also note the soft lights providing general illumination off-camera to the left and right. The skull itself contains a cool white pico but is illuminated from above with a blue key light to give it a soft glow.

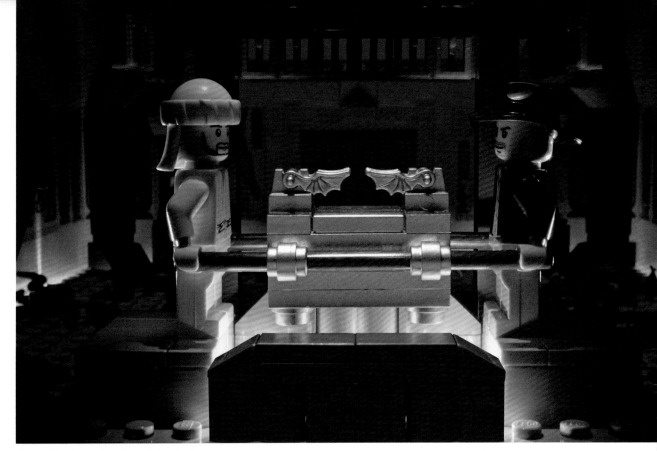

Figure 3-18: *Finding the Ark* (author)

Figure 3-19: *Interrogation Tent* (author)

Figure 3-20: *Men in Brick Secret Armory* (author)

MEN IN BRICK SECRET ARMORY

The scene in Figure 3-20 shows two government agents retrieving futuristic firearms from a secret armory hidden behind the walls of an unsuspecting family's apartment. The scene contains 24 separate 1×4 plate lights carefully placed as backlights to provide soft, diffuse light. The walls are made from trans-clear plates with an illuminated white wall behind (see Figure 3-21). The ceiling lights are light-diffusing gels backlit with the same 1×4 down lights. The table lamps are made using pico LEDs. The only other light comes from two more down lights providing fill light above the coffee table. Note that the wires are connected using old 4.5 V connectors.

Figure 3-21: *Men in Brick Secret Armory* exterior view

THE BATCAVE

Large caverns present a great opportunity for lighting. In Figure 3-22, under-cabinet strip lighting, placed behind rocks on the ceiling and angled toward the back, illuminates the rocks' texture. In doing so, it also backlights stalactites, stalagmites, and a large family of bats hanging about. A combination of carefully placed key lights then illuminates the major areas of activity and cool vehicles. The bricks may make this build, but the lighting makes the atmosphere.

Figure 3-22: *The Batcave* (Wayne Hussey and Carlyle Livingston II)

PROJECT 3: BUILD YOUR OWN LED STUDIO LIGHT

Now it's your turn to experiment with lighting by building your own miniature studio light (see Figure 3-23). This build utilizes your own bricks, plus a common pico LED available from several manufacturers, along with a discontinued LEGO 9 V light reflector dish (part #x85 on BrickLink). Pull the handle on the rear to adjust the projected light pattern from wide to spot. Adjust the barn doors to block light from falling where it doesn't belong. Note that

Figure 3-6 shows a smaller key light (lower right) built using the same technique.

Requirements

- One pico LED, available from Brickstuff or equivalent third-party lighting manufacturers
- One adapter board with at least two connectors, as required to connect the LED to the battery pack
- One battery pack or other source of power

Figure 3-23: Project 3: LED studio light

1x
85861
Black

1x
4589b
Black

3x
32192
Black

1x
60474
Light Bluish Gray

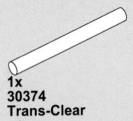

1x
30374
Trans-Clear

2x
2780
Black

3x
32039
Black

1x
57585
Black

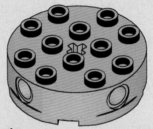

1x
6222
Light Bluish Gray

1x
x85
Chrome Silver

2x
32013
Black

4x
3705
Black

4x
35480
Blue

6x
2540
Black

3x
32073
Black

4x
27925
Blue

6x
2335
Black

1x
11833
Light Bluish Gray

2x
18674
Blue

1

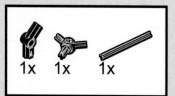

1x   1x   1x

2

2x   2x

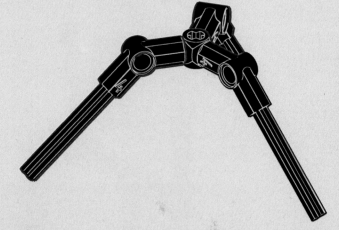

3

3x   2x

4

2x   2x   2x

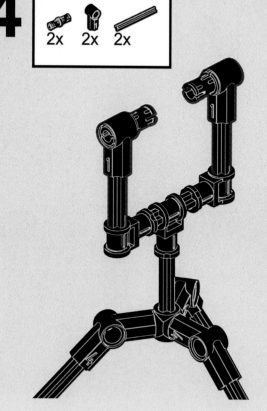

5

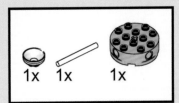

1x 1x 1x

 6

4x 1x

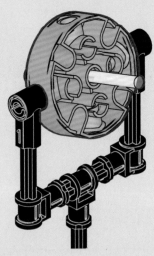

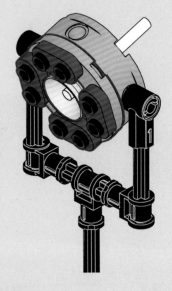

7

6x

8

6x

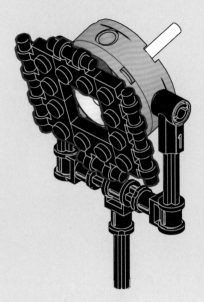

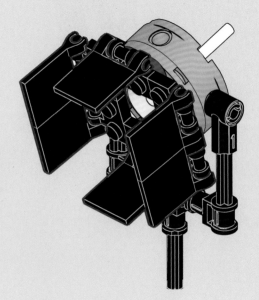

9

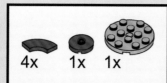

4x 1x 1x

10

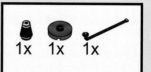

1x 1x 1x

Insert LED into cone.

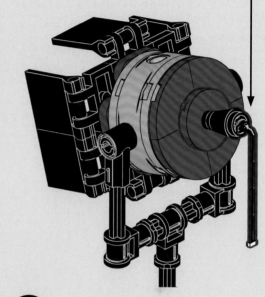

11

1x

Thread 1×1 round plate over LED and attach.

12

Pull/push cone to adjust spotlight.

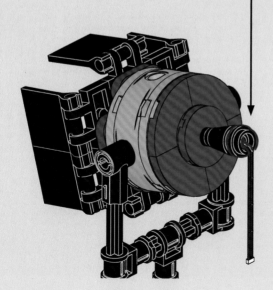

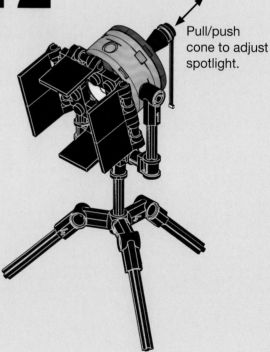

4 EXTERIOR LIGHTING

More and more, so it seems to me, light is the beautifier of the building.

—FRANK LLOYD WRIGHT, ARCHITECT

Ever since Thomas Edison used his incandescent bulb to light a street in Menlo Park, New Jersey, on New Year's Eve 1879, electric light has been a ubiquitous part of the urban landscape. Architects, like the public, were quickly drawn to the creative possibilities of artificial lighting, such that exterior lighting design is now a key element of modern buildings, parks, and public spaces. For LEGO builders, too, exterior scenes provide a range of opportunities to incorporate lighting.

Celebricktion (Harry and Austin Nijenkamp). This magnificent LEGO city was lit using over 1,000 Brickstuff LEDs. The 225 baseplates are affixed to 25 large transport panels that conceal the wires.

ARCHITECTURAL LIGHTING

LEGO buildings, like their real-world counter-parts, offer numerous opportunities for lighting not just interiors but exteriors as well. If it's fun to assemble a whole street of LEGO buildings, it's even more fun to see them lit up, as in Figure 4-1.

Architectural lighting incorporates light as a design element to enhance a building's visual aesthetic, its functional usability, and the view-er's experience of the space. Light becomes another resource, just like stone, steel, or glass, and its use can range from the subtle to

Figure 4-1: A lit LEGO street display (Jim Pirzyk)

the sublime. In this section, we'll explore four lighting techniques that will make your LEGO city really shine: window lighting, facade lighting, landscape lighting, and street lighting. Lastly, we'll discuss a key consideration of scaling up lighting from a single building to a city: managing wires.

WINDOW LIGHTING

At night, the patterns of light that project outward through a building's windows can define the building's contours and character. There are a few ways to achieve this *window lighting* effect. If the interiors of your buildings feature realistic detail, light them following the techniques discussed in Chapter 3 and simply allow that light to shine through the windows. LEGO Café Corner buildings, for example, typically come with elaborate interiors. They can be expensive to light, however, since details add cost, weight, and fiddly bits that can break off when you move the building. Therefore, it's best to leave fully detailed interiors for just your favorite foreground buildings that visitors can inspect close up.

If a building is far enough back in your display that viewers won't be able to see anything through the windows, you can forgo the interior details. Buildings used to form a city skyline, as in Figure 4-2, can even be built as "shell buildings" without floors. This way the light from a few well-placed LEDs can filter through the entire structure. When you light buildings using this approach,

it's important that the bricks that show on the inside be a light color so they'll be visible when lit. Alternately, place translucent light-diffusing window film (such as that made by Woodland Scenics) behind each window to create the impression of interior rooms without the details. This approach is commonly used in the model railroad hobby and works very well for background structures where interior details won't be seen. You can even place pieces of opaque tape on the diffusion film to simulate window shades and blinds. For a pure LEGO solution, skip the film and build windows using trans-clear plates.

Sometimes a pure LEGO solution is very effective. The tall black building on the far right in Figure 4-2 uses a mixture of trans-clear and trans-black panels to vary the light from a single string of Christmas lights hung inside the building from the top. From a distance, this technique produces a remarkably realistic effect.

REDUCING LIGHT BLEED

Some LEGO colors aren't very opaque and permit light to bleed through an element, especially when a bright LED is placed up against it. To alleviate this problem, install an opaque sticker on the inner face of the element before attaching the LED. Opaque stickers can be made from adhesive automotive muffler tape or cut from dark vinyl colors.

Figure 4-2: *A LEGO City Skyline* (Roger Snow)

FACADE LIGHTING

Facade lighting involves the careful placement of exterior lights to highlight architectural lines, details, and features of a building. Some third-party manufacturers provide kits for lighting the facades of popular LEGO sets. These kits typically feature many bright LEDs, but may not reflect the actual lighting plans of their real-world counterparts. If capturing the realistic lighting plan of a recognizable subject is your goal, then compare the available kits carefully. The model of the United States Capitol shown in Figure 4-3 is a good example

Figure 4-3: *United States Capitol* (Wayne Tyler)

of a custom build with lighting that closely follows that of its namesake. Notice how the vertical columns on the building are backlit, while the dome is illuminated with four focused lights, just like the real Capitol.

LANDSCAPE LIGHTING

Not all exterior LEGO creations center on buildings; some feature landscapes, too. *Landscape lighting* uses lights to define the contours of tree lines, walking paths, fountains, and other landscaping elements. To create realistic landscape lighting, illuminate paths that people follow and major elements of park designs. For example, try installing spotlights underneath trees and along paths. Use dimmers so park lights don't overwhelm the other lights in an area. The Capitol model

in Figure 4-3 features a thoroughly lit landscape in the foreground. For another great example of landscape lighting, see Figure 4-7 later in the chapter.

STREET LIGHTING

Large cityscapes or designs that feature street scenes typically incorporate *street lighting* for realistic illumination. Several third-party manufacturers sell prelit lampposts for this purpose and/or LEDs specially made for use with LEGO's lamppost element (see Figure 4-4). Some of these lamppost designs have the LED facing upward, however, which projects most of the light away from the street or sidewalk. For a more realistic solution, use a pico LED and mount it upside down in the top of the lamp reflector (usually a 2×2 radar dish).

Figure 4-4: A street lamp (Brickstuff)

MANAGING WIRES

One challenge of lighting a display of buildings is organizing and hiding all the wires. The solution you choose will depend largely on how portable and flexible your display needs to be. If it's permanently installed at home and won't be traveling, you can run the wires and connectors out the backs of the buildings. Then plug the wires into a set of distribution points, which can range from tiny plug boards to large USB hubs, depending on the manufacturer's system that you're using. You can mix systems as needed for each individual building. If your display will be traveling to shows, however, or if you need flexibility to

rearrange buildings, as in a collaborative club display with different modules, then consider standardizing the lighting connection interfaces to a single manufacturer's plug type.

MULTI-CIRCUIT DESIGN

LEGO buildings and cityscapes are built modularly, and it helps to think in the same terms when you're planning the lighting design. For convenience, foreground buildings that you want to be able to open up to show off the interior should use a single circuit of LEDs with one power cable for each floor. Run the connector end of each power cable out the back of the floor and down to ground level, where you can connect all the floors' cables to a central power hub. This way, you can easily take the building apart floor by floor to reveal the interior; all you have to do is unplug a single cable per floor at the ground-level hub. Shell or background buildings that have little or no interior detail to show off can be wired together on a single circuit.

Remember not to attach more LEDs to a power hub than its power supply can support (see Chapter 2). If the current need is too high,

split your buildings' power cables across multiple power hubs conveniently located along the back of a row of buildings or under buildings in the middle of a display. It helps to start with a wiring diagram, which you should keep for future reference.

CUSTOM WIRING

For larger and more portable displays, it may be advantageous to build your own power and wiring system, as in the collaborative model space station shown in Figure 4-5. This display features modular interior rooms that are

Figure 4-5: *The Space Station of Doom* (Kenosha LEGO User Group/ John Wolfe)

lighted just like the floors in a row of buildings. The custom barrel connectors, thick wires, and high-capacity power supply allow a large number of modules to be lighted.

This option is definitely DIY and requires soldering skills, but it can be very cost-effective. Plus, standardized connectors make for quick setup and can be easily reconfigured to accept however many modules are needed. This gives you maximum flexibility to rearrange and light the modules.

CONNECTING FLOORS AND BASEPLATES

Using a single circuit and plug per floor can lead to a tangle of cables if you have a lot of buildings to light. Fortunately, third-party LEGO lighting manufacturers make special connectors to magically connect floors and baseplates together. Figure 4-6 shows some examples.

Vertical floor connectors like the example on the right allow you to electrically connect two floors together by simply setting one on top of another. This makes it very easy to disassemble a building to show off the interior and then quickly reassemble it. Likewise, there are horizontal electrical connectors, such as the left two examples, that can electrically connect two baseplates together, greatly simplifying the wiring for larger city displays.

Figure 4-6: Third-party floor and baseplate connectors

EXAMPLES OF CREATIVE EXTERIOR LIGHTING

Now that you've learned a bit about exterior lighting styles, let's take a look at how you can combine them to make some amazing MOCs. Study the way that light plays a key role in each of these scenes. With a little practice and a lot of patience, you'll soon be creating citywide scenes like these, too.

NATIONAL MALL

Figure 4-7 shows a microscale version of the Washington Monument and National Mall area in Washington, DC. The artist uses tiny 3 mm bars standing on end to represent streetlights. Other lampposts use trans-clear faceted rock/jewel elements (#30153) for the lampshade on top (see Figure 4-4 for another view of these). Notice how the streets and walking paths are defined with path lighting.

Figure 4-7: *National Mall* (Wayne Tyler)

KENT THEATRE

The fantastic theater in Figure 4-8 is packed with custom lighting. The marquee is backlit with black and trans-clear bricks, and the artist has skillfully strung EL wire around it to simulate glowing neon lights. The streetlamp and carefully placed stage entrance lamp on the far right make the scene come alive. Even the custom-printed movie posters are mounted on trans-clear bricks and backlit.

Figure 4-8: *The Kent Theatre* (Jarren Harkema)

HOGWARTS

The legendary Hogwarts School of Witchcraft and Wizardry has been rendered many times in brick, but Figure 4-9 shows one of the very best versions when it comes to lighting. Notice how the window lights define the castle's shape from a distance. Meanwhile, the lower rocks are illuminated by cool white lights to contrast with the warm white lights coming through the castle windows.

Figure 4-9: *Hogwarts* (Hyungmin Park)

CYBERPOCALYPSE

The dystopian vision of the future shown in Figure 4-10 is a medley of building and lighting techniques. Individual LEDs are peppered throughout the five-foot-high buildings, along with various colors of EL wire. Some buildings are lit only through window lighting, while others feature facade lights to define their forms. All these contribute to the bustling, futuristic atmosphere.

Figure 4-10: *Cyberpocalypse* (Carter Baldwin, Nate Brill, Kyle Vreze, Forest King, Ignacio Bernaldez, Sam Wormuth, Alex Valentino, and Chris Edwards)

VILLA AMANZI

The striking residence seen in Figure 4-11 is a prime example of window lighting. The horizontal lines of the structure are exclusively defined by the interior lights coming through the expansive windows. The same light plays on the other parts of the structure and the surrounding landscape.

Figure 4-11: *Villa Amanzi* (Robert Turner)

Figure 4-12: *Airport* (Chris Rosek)

AIRPORT

Airports have a lot of opportunities for lighting, and the scene in Figure 4-12 takes
advantage of most of them. The concourse,
control tower, and service areas are brightly
illuminated. Several aircraft have interior,
taxi, and navigation lights. And the runway
lights are a perfect example of the benefits
of daisy-chaining lights to prevent long and
complicated cable runs.

PROJECT 4: BUILD YOUR OWN SKY SIGNAL

Now it's your turn to experiment with exterior lighting by building your own sky signal (see Figure 4-13), a searchlight that projects a focused image, commonly called a Leko. The build uses a bright LED and a LEGO magnifying lens to project the image from a transparent 1×2×2 panel (BrickLink #4864bpb039, a rose) from about 12 inches away.

If you'd prefer to project a different image with your sky signal, search BrickLink for trans-clear 1×2×2 panels decorated with other patterns. Alternatively, make your own custom image using a vinyl cutter, or print one onto clear transparency paper, and stick it onto a trans-clear 1×2×2 panel. If you have a steady hand, you can even apply Scotch Magic Invisible tape to the panel and use a fine-line permanent marker to hand-draw a custom image. Perhaps even an image of . . . a bat.

Use this lamp to paint images onto building facades or light up the sky with a special logo or message. Keep in mind that the image projects upside down and backward due to the lens.

Requirements

- One high-brightness LED, such as a 1 watt LED from Brickstuff or a high-brightness Bit Light from Light My Bricks
- Adapter boards and wires as needed to connect the LED to the power supply
- One power supply or battery pack
- Custom pattern for projection or a decorated 1×2×2 LEGO panel
- Invisible tape (optional)

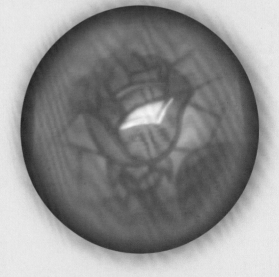

Figure 4-13: Project 4: Sky signal

2x
43093
Black

2x
37352
Black

4x
32064
Black

1x
30658
Black

3x
15573
Dark Bluish Gray

4x
87087
Light Bluish Gray

2x
3700
Light Bluish Gray

1x
60592
White

1x
3023
Dark Bluish Gray

1x
2654
Light Bluish Gray

1x
30152c03
Black

1x
14769
Dark Bluish Gray

4x
27925
Light Bluish Gray

1x
60601
Trans-Clear

2x
2825
Black

1x
32802
Dark Bluish Gray

1x
60474
Light Bluish Gray

1x
32073
Black

1x
3020
Dark Bluish Gray

1x
11833
Light Bluish Gray

1x
87552
Trans-Clear

2x
4871
Dark Bluish Gray

1x
61485
Black

2x
3005
Light Bluish Gray

2x
30987
Light Bluish Gray

1

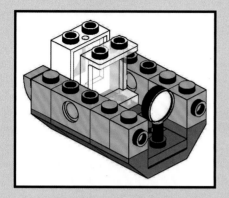

2

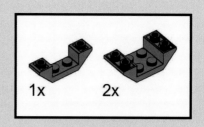

1x 2x

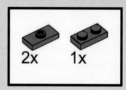

3

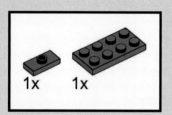

1x 1x

4

2x 1x

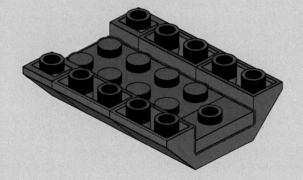

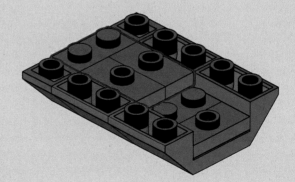

5

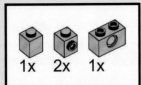

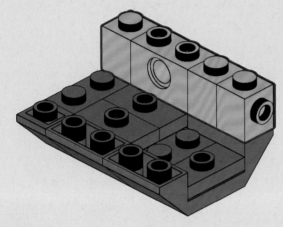

6

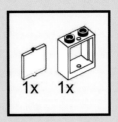

Light diffuser

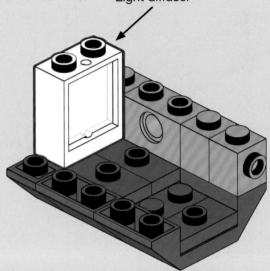

7

Image

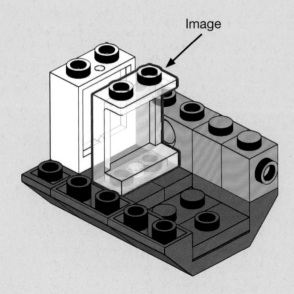

8

Lens

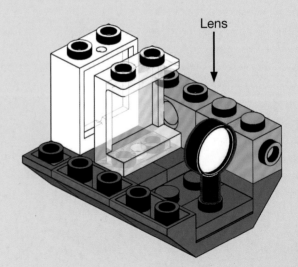

9

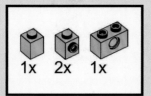

1x 2x 1x

10

2x

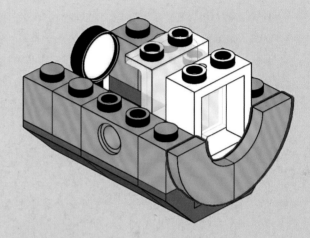

11

12

13

1x

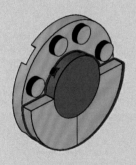

14

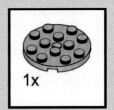
1x

15

2x   1x

16

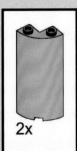
2x

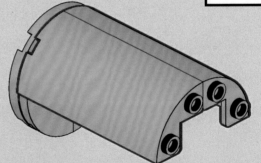

17

1x

18

1x

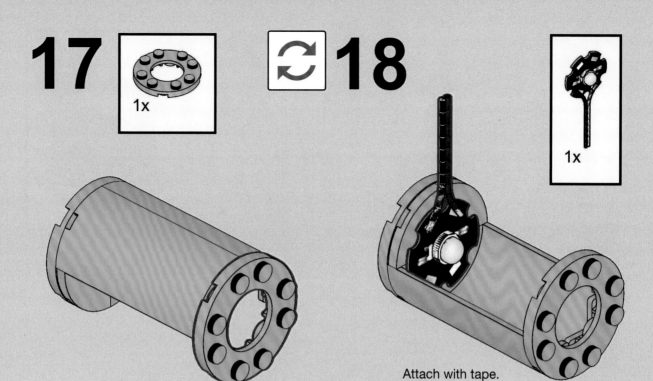

Attach with tape.

19

20

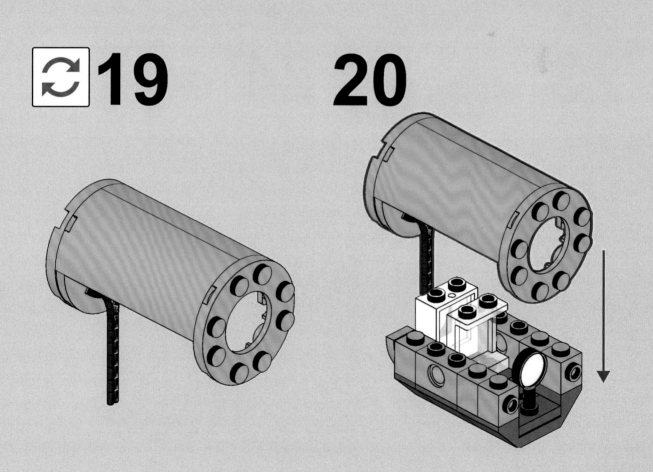

21

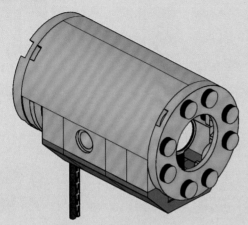

22

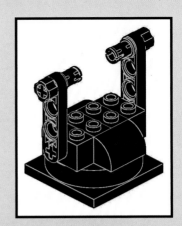

23

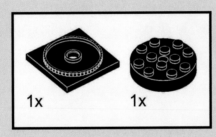

1x   1x

24

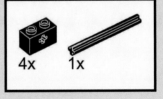

4x   1x

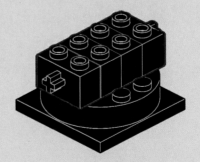

25

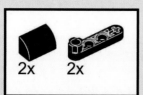

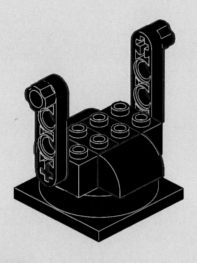

26

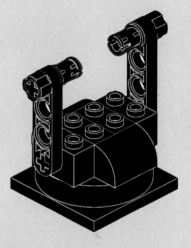

27

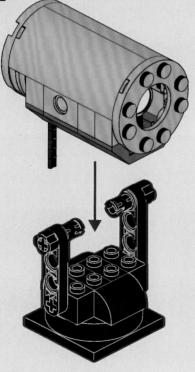

28

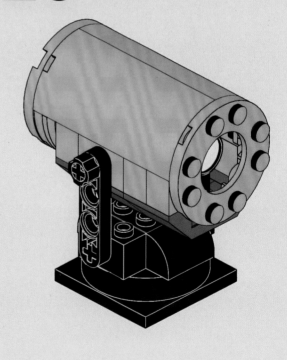

5 LIGHTING WITH CHARACTER

In a dark place we find ourselves, and a little more knowledge lights our way.

—YODA, JEDI MASTER

Lighting can be very personal. Sometimes we associate a specific style of lighting with a particular person or character so intensely that the light itself becomes part of the character. For example, mentioning the sinister Darth Vader immediately conjures up images of a flaming red lightsaber, while you can't think of Iron Man without picturing the glowing arc reactor in his chest. Miners deep underground naturally have yellow-tinted lamps on their helmets, and a night watchman just has to be carrying a bright lantern or flashlight. Adding these sorts of lights to LEGO characters adds interest to both the individual characters and to the larger scenes that they inhabit. This chapter will show you how to use lighting to make your LEGO characters come alive.

A Little Light Swordplay (author). This vignette uses light both to create a threatening mood and to draw the viewer into the action. Several third-party lighting suppliers sell prelit lightsaber bars. Notice how much interest the lightsabers add to this scene.

LIGHTING MINIFIGURES

So popular are LEGO minifigs that AFOLs refer to builds scaled to the little figures as being at *minifig scale*. Because minifigs are so ubiquitous, lighting them is a common customization. You can use prelit minifigs and accessories or wire up your own accessories with custom LED lighting. The latter is a true DIY endeavor and can be tricky, involving cutting or drilling into the minifigs. However, recent advances in wireless LED technology can greatly simplify the process.

PRELIT MINIFIGS

There aren't many all-LEGO solutions for lighting minifigs available to the LEGO purist, but LEGO did make several "Light-Up Lightsaber" *Star Wars* minifigs in 2005. Outwardly, these appear to be normal minifigs, except for the slightly larger lightsaber hilt. However, inside the minifig a small battery and LED are hidden. Pressing down on the minifig's head makes the lightsaber glow. LEGO used the same design for a police minifig carrying a flashlight. These Light-Up minifigs can still be purchased through BrickLink.

Today, several third-party lighting companies sell prewired lightsabers that you can add to your minifigs. Occasionally you can find lit flashlights, lanterns, and other handheld accessories as well. Some examples are shown in Figure 5-1.

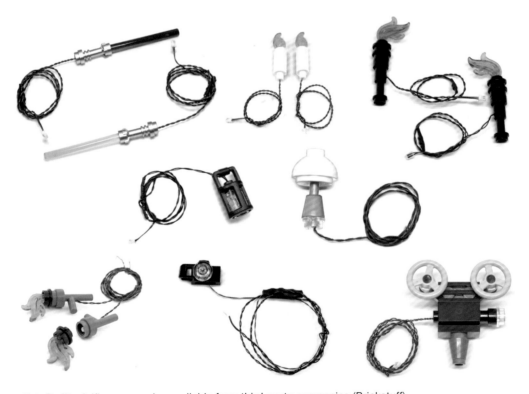

Figure 5-1: Prelit minifig accessories available from third-party companies (Brickstuff)

Some third-party manufacturers, like Eclipse-GRAFX, custom-print LEGO elements representing televisions, control panels, heads-up displays, stained glass windows, and other accessories that minifigs can interact with. These sorts of accessories can be lit as well, as shown in Figure 5-2.

Some third-party lighting manufacturers, like Brickstuff and LifeLites, also do custom work. If you need to add lighting to a custom minifig or an unusual accessory, you can reach out to one of these companies and ask if they can help with your project. Keep in mind that some minifig accessories, like torches,

cameras, and weapons, can look even better when animated with a lighting effect controller (see Chapter 9).

DIY MINIFIG LIGHTING

If you want to light your own minifigs and handheld minifig accessories, the main question is what to do with the wires. The most effective solution is to conceal the wires inside the minifig itself and run them out the back of the minifig's legs. You can accomplish this in three easy steps:

1. Drill holes to create a path for the wires.

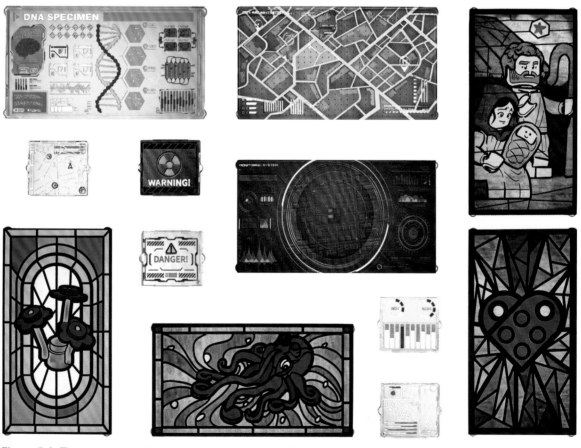

Figure 5-2: These custom-printed window glass elements from EclipseGRAFX can be backlit to create detailed heads-up control panels, computer screens, and stained glass windows.

2. Thread the LED wires through the holes.

3. Secure the wires into place.

Figure 5-3 shows an overview of the process. You'll need a miniature drill or pin vise with a 2 mm drill bit, as well as some urethane or white glue—materials found at most hobby shops. You can also use a motor tool, like a Dremel, for the drilling, but be sure to set it to the lowest speed so that you don't melt the plastic.

Drill Holes

To create a custom-lit minifig with hand-held accessory, the first step is to drill holes through the minifig arm, hand, and legs, as well as through the accessory. Figure 5-3 outlines the process. First, remove the minifig's arm and hand from its torso. Then, drill halfway into the arm through the center of the shoulder connector (a). Don't drill all the way through. Next, drill into the arm through the

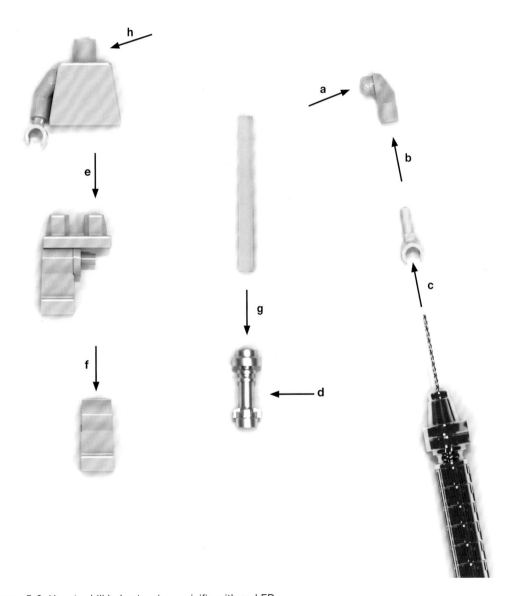

Figure 5-3: How to drill holes to wire a minifig with an LED

hand socket (b). Drill straight back, but don't go all the way through the elbow. You should hit an air bubble caused by the molding process.

Now you can move on to the hand and the accessory. Drill clear through the center of the connector pin on the hand (c). Then, find the spot on the side or back of the accessory where the minifig will hold it and drill a hole there as well (d). Make sure the hole aligns with the hole in the hand so the wires can pass from the accessory into the hand.

Finally, turn your attention to the minifig's legs. Drill through the center of one of the hip studs (e), going straight down and through the leg (f). There are two studs, so make sure to drill the hole through the stud on the side of the torso with the arm holding the lighted accessory.

Thread the Wires

The next step is to thread the wires through the holes that you've drilled. Start by choosing an LED small enough to fit into whatever accessory you're lighting. Most likely this will need to fit through a hole drilled into a 3 mm bar. You can solder the wires on yourself, but soldering LEDs this tiny is very tedious work, so consider purchasing prewired LEDs. If you have a prewired LED with a connector and neither end is small enough to thread through the holes you've drilled, then cut the wire about two inches from the connector end. After you've threaded the wire through the figure, you can solder the connector end back on.

To wire the figure, start by teasing the wires through the lighted accessory (g and d in Figure 5-3), then the hand (c), then the arm. Push the wires through the arm, entering at the wrist (b). The wires should bend upward once they hit the elbow to get up to the shoulder. Then, pull the wires out through the shoulder (a), using a toothpick, dental tool, or fine tweezers if needed. Next, thread the wires into the torso (h), out the bottom, and through the holes in the hip (e) and leg (f). Finally, pull the wires out the hole on the back of the leg. Alternatively, you can drill a hole into the stud the leg will attach to and pull the wire through it to completely hide the wire from the scene.

Once you've threaded the wires, get the LED into its final position and pull the wires tight through the accessory. Then, gently snap the accessory into the hand and the hand into the arm, but don't snap the arm into the body just yet.

Secure the Wires

Before securing the wires, test the LED to make sure it works. If it lights, place a bead of glue into the hole on the shoulder of the arm piece (a), where the wires are coming out, to hold the wires in place. Once the glue is dry, snap the arm into the shoulder socket (h). Then slide the torso down onto the hips (e) while gently pulling the remaining slack on the wire down through the leg (f). Now you're ready to connect your minifig to a power supply and place it in the scene of your choice.

You can also use the steps just outlined to install LEDs anywhere inside a minifig itself.

For example, you could install an LED inside a minifig's head to illuminate its eyes or inside its torso to light up a chest device. In either case, you'd want to drill a small hole wherever you want the light to come through. LEDs can also be used to illuminate translucent body parts. As of this writing, LEGO hasn't made all the elements of a minifig in translucent colors. However, fully translucent minifig clones are available online through some third-party manufacturers.

WIRELESS LIGHTING

The same technology that wirelessly charges your cell phone can also be used to wirelessly transmit power to LEDs. The equipment, commonly listed as *inductive charging LEDs*, provides a convenient way to add LEDs to minifigs (and vehicles, too) without the need for wiring, giving you freedom to move them around the scene or place them in positions where wires would be impossible to hide. You can find inductive LED kits on eBay or through specialty electronics distributers such as Adafruit, though the options are limited and more expensive than wired LED installations.

Inductive lighting systems are composed of two parts: a coil to transmit the power through the air and one or more LEDs with small coils to receive the power (see Figure 5-4). The LEDs will light anywhere from a couple centimeters to dozens of centimeters away from the coil, depending on the coil size and the power used. Wireless LEDs are definitely DIY, requiring a basic knowledge of electronics to use correctly.

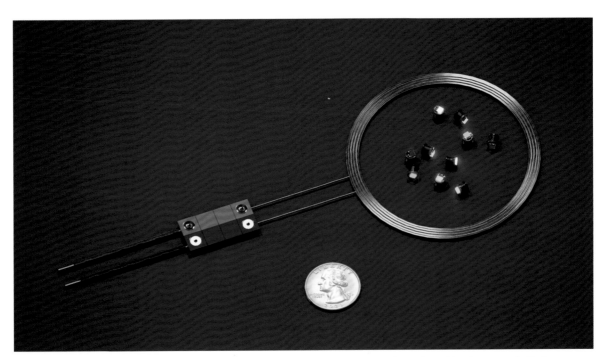

Figure 5-4: Small enough to fit inside a 1×1 brick, wireless LEDs illuminate when placed in close proximity to an energizing coil.

Figure 5-5: *Cornered* (author)

EXAMPLES OF CHARACTER LIGHTING

Now that you've learned how minifigs can be wired for lights, let's take a look at some great examples of custom character lighting. With a little experimentation, you, too, can make your figures really shine and add focus and interest to your LEGO scenes.

CORNERED

The scene in Figure 5-5 uses custom mini-fig lighting to convey a sinister, shadowy mood as a group of gangsters approaches. Note the tiny LEDs wired into the gun barrels held by the figures on the left and right.

Meanwhile, LEDs inside the custom cigarette and lighter cast an orange glow on the center figure's face. The guns are animated using a lighting effect controller (more on LECs in Chapter 9).

Figure 5-6: *Ulric the Firebright* (Cornbuilder)

ULRIC THE FIREBRIGHT

While this chapter has focused on lighting minifigs, it's also possible to light larger-scale characters. These can range in size from Miniland scale (the size of the brick-built figures used in the Miniland sections of LEGO theme parks) to the buildable figures in LEGO's own product line. The large character in Figure 5-6 is similar in size to a buildable LEGO figure. LEDs have been carefully positioned behind trans-colored bricks in the chest and head to give forth an internal glow.

IRON GUY

Thanks to inductive charging LEDs, the customized flying minifig shown in Figure 5-8 is lit in midair, without the need for wires. The figure is mounted on a trans-clear bar suspended above the ground. Six wireless LEDs provide the glowing light in the head, torso, hands, and feet. Some cutting and drilling was needed to fit the LEDs inside the head and torso. There are actually three inductive LEDs inside the torso. One lights the arc reactor. The other two have had their LEDs removed so just the coils remain. Very thin wires were then soldered to these coils and threaded through the arms and hands to light the LEDs in the hands, which have been placed inside a short piece of trans-clear 3 mm bar.

Figure 5-7: *Iron Guy* (author)

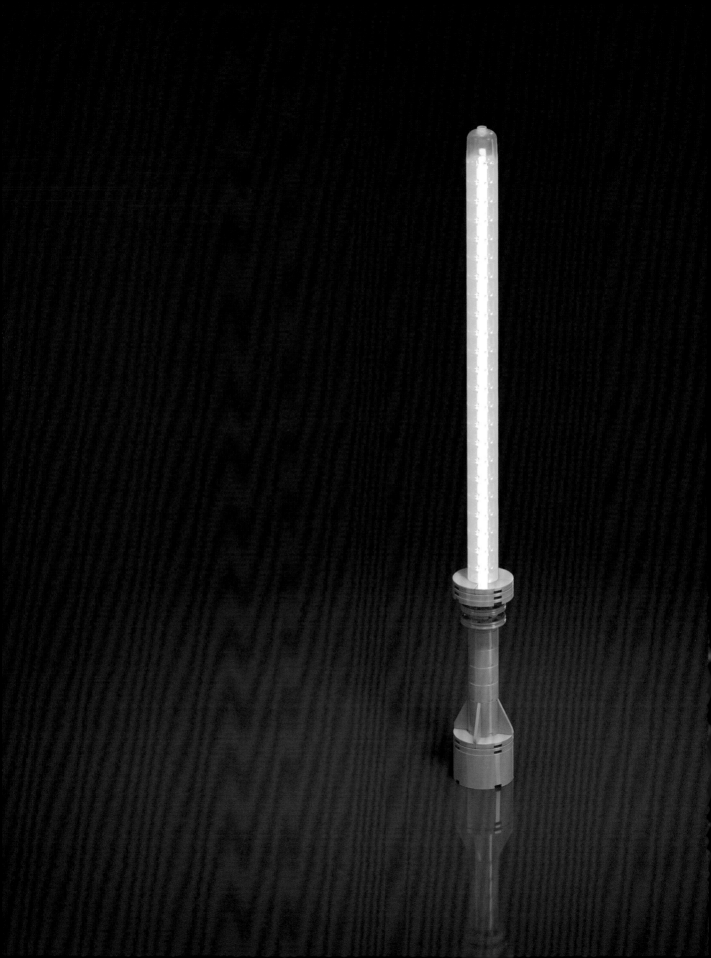

PROJECT 5: BUILD A LIGHT SWORD

Now it's your turn to experiment with figure lighting by building a large light sword (see Figure 5-8). It's most decidedly swooshable, and it can be used with the LEGO *Star Wars* buildable figures. This build utilizes two LED Light Bars from Light My Bricks, but you can substitute a Brickstuff Flexible LED Filament if desired. Alternatively, you can use several pico LEDs available from various third-party manufacturers, although these won't produce as smooth a result. Try building several light swords in different colors to create your own space battle. For best results, view the light sword in a dark room.

Requirements

- Two 12 cm Light My Bricks LED Light Bars or a Brickstuff 300 mm Flexible LED Filament. For the latter, you'll need to solder an appropriate resistor or contact the manufacturer to do this for you.
- Adapter boards and wires as needed to connect the LED to the power supply.
- One power supply. The space in the bottom of the hilt can accept a small coin-cell battery pack to make this build untethered, or you can run a wire out of the hilt to a larger battery pack or other power supply.

Figure 5-8: Project 5: Light sword

1x
4591
Light Gray

3x
60474
Light Bluish Gray

23x
3941
Trans-Light Blue

8x
27925
Light Bluish Gray

2x
4032b
Light Bluish Gray

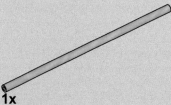

1x
75c11
Light Bluish Gray

1x
30151a
Trans-Light Blue

2x
15535
Light Bluish Gray

4x
3941
Light Bluish Gray

2x
4185
Light Bluish Gray

2x
24593
Light Bluish Gray

1

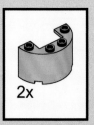

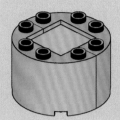

2x

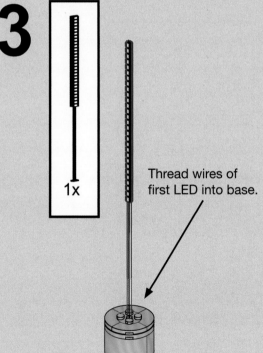

2

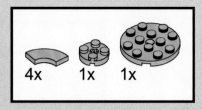

4x 1x 1x

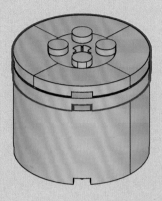

3

1x

Thread wires of
first LED into base.

4

1x

Thread wires of
first LED alongside
second LED into base.

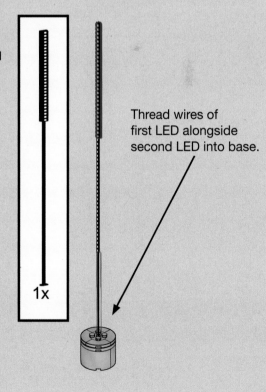

5

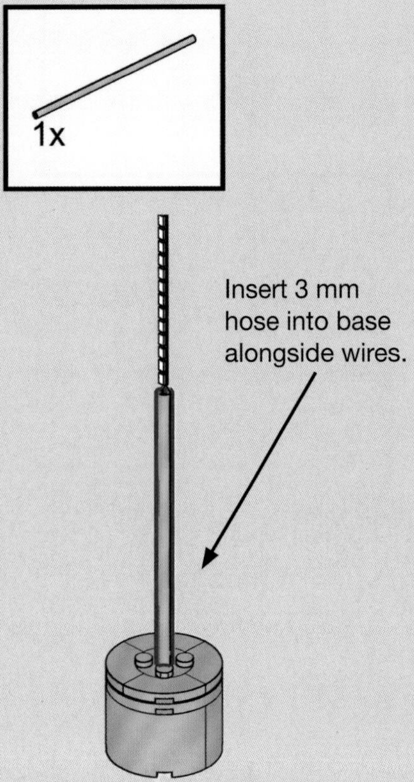

1x

Insert 3 mm hose into base alongside wires.

6

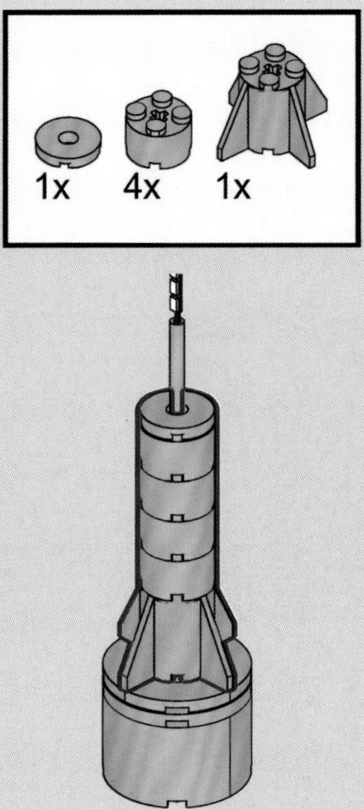

1x   4x   1x

7

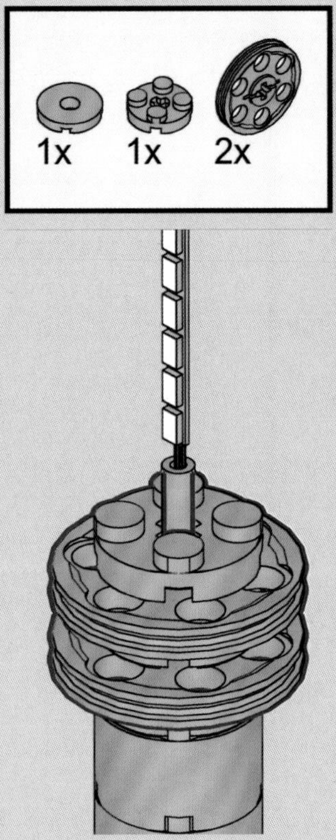

1x   1x   2x

8

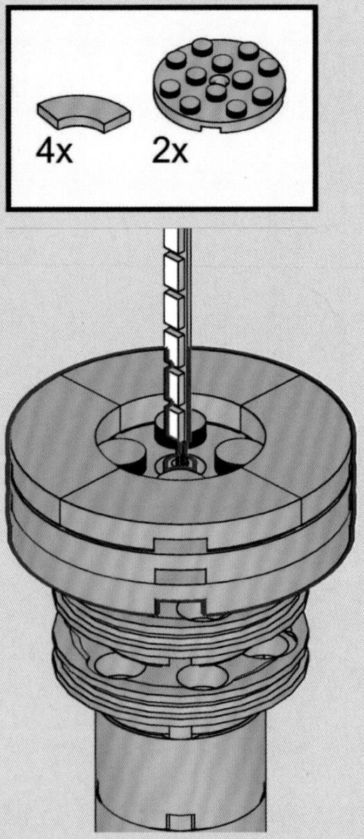

4x   2x

9 2x

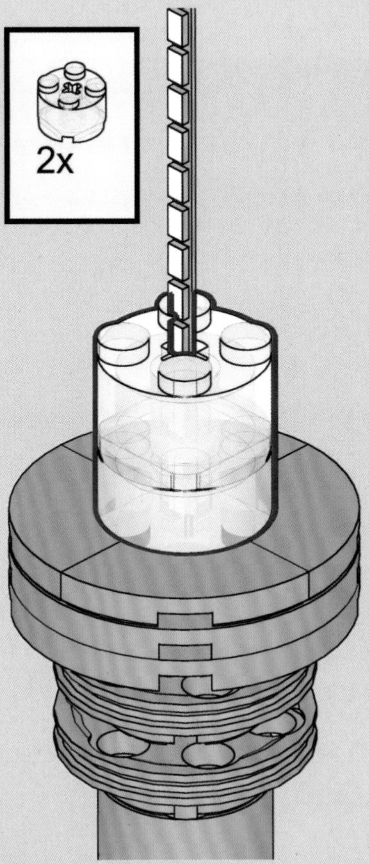

10 10x

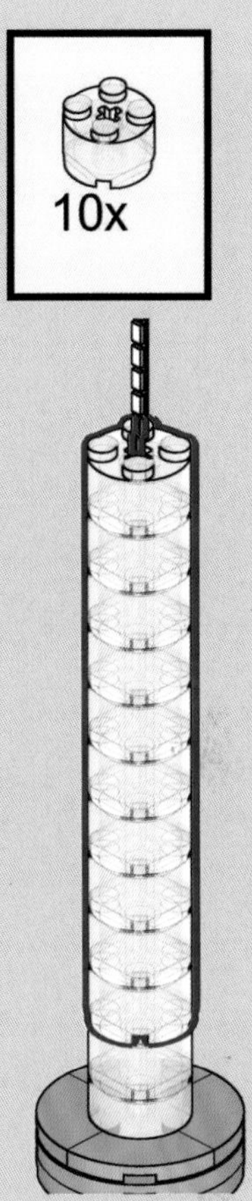

11 11x 1x

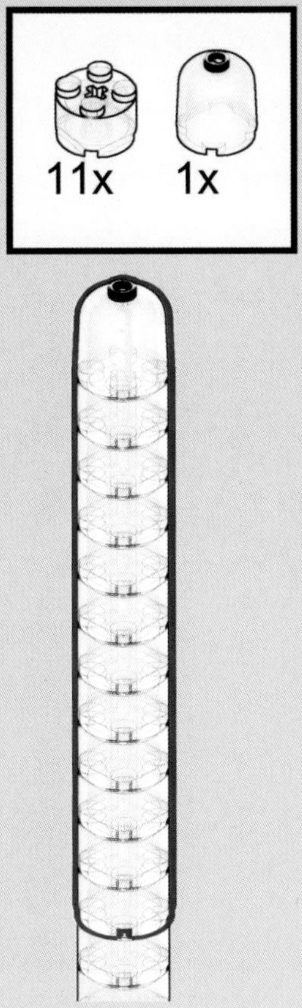

12

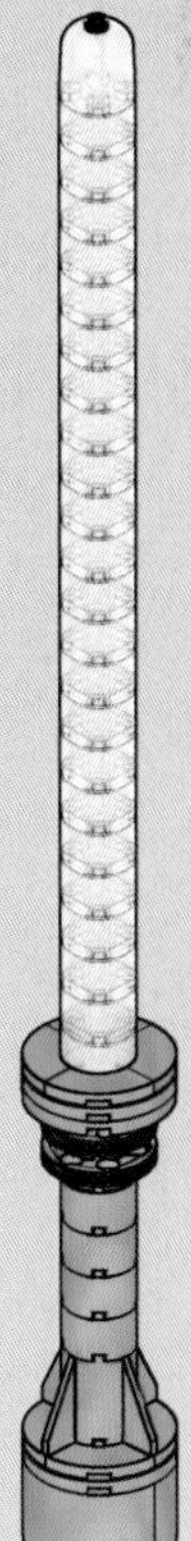

6 VEHICLE LIGHTING

Why don't you fix your little problem and light this candle.

—ALAN B. SHEPARD, ASTRONAUT

Vehicles past, present, and future are a perennial favorite of toy makers, including the LEGO Group. Beyond their speed and power, vehicles offer a variety of unique lighting opportunities, including headlights, navigation lights, and engine lights. Each lighting technology through the years has had its own distinctive look: oil lamps and tungsten electric lamps give off a yellow glow, while fluorescent bulbs have a blue-green cast and arc lights and modern halogens are whiter. Accurately capturing the lighting of a prototype vehicle of any era can make a good MOC really stand out.

Union Pacific EMD SD70 Ace Locomotive (Dennis Glaasker). This supersized locomotive is built at 1/16 scale and measures a whopping 48 inches long, more than three times the size of a typical LEGO locomotive. It features a wealth of prototypical details, including cabin, navigation, and ground lights.

VEHICLE LIGHTING TECHNIQUES

LEGO vehicle lighting utilizes many of the same components and techniques employed in lighting buildings and other brick creations. However, while many general-purpose lighting techniques apply, there are a few vehicle-specific details that merit special discussion, including headlights and lightbars. Additionally, because LEGO vehicles are often small and are expected to be mobile, they come with their own particular battery and wiring considerations.

HEADLIGHTS

Many years ago, LEGO created a dedicated element for headlights on cars, popularly known as the *headlight brick* (BrickLink #4070). Not only does this brick make great-looking car headlights, but it works perfectly for actually lighting them too! Because

the brick is open on the back, you can slip a 3 mm through-hole LED through the back and into the circular opening on the front, as shown in the center example of Figure 6-1. You can then add a translucent plate to the front to color the light.

Other 1×1 LEGO bricks can also work as headlights. The left example in Figure 6-1 shows a pico LED inserted through the front of a 1×1 brick with stud on one side (BrickLink #87087). The example on the right shows a 5 mm through-hole LED inserted through a 1×1 Technic brick (BrickLink #6541) with a trans-clear 1×1 round plate inserted stud-first into the opposite side. Note that any element with a clip can also hold a 3 mm through-hole LED, while a 5 mm through-hole LED will usually fit into the open stud-size holes of Technic liftarms, connectors, and other elements.

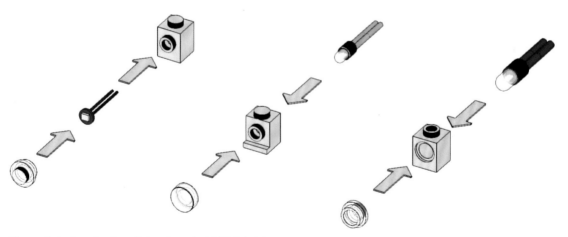

Figure 6-1: Creating headlights from 1×1 LEGO bricks

LIGHTBARS

Emergency vehicles often have a lightbar on the roof that flickers and flashes. LEGO offers purists a few official options, including the Light & Sound elements made in the late 1980s (see left example in Figure 6-2) and the Light & Sound siren from 2019 (the center example in Figure 6-2). The former has a few disadvantages. While its lights do flash, they don't always flash in sync with each other. Also, if the lighting needs to be self-contained, you'll need space in the vehicle for a large (4×8×3) 9 V battery box. The latter solution is more compact: 2×6×1, with battery included. However, the siren only runs for 5 seconds with each press, and the LEDs aren't very bright.

Enter the third-party lighting manufacturers. LifeLites makes a product called a StudLite that features tiny colored LEDs surface-mounted to a thin, flexible ribbon designed to fit on top of a 1× plate and wrap around back and underneath (see the right example in Figure 6-2). This lets you place a translucent 1× round plate or tile on top of the StudLite and build the lightbar into a vehicle's roof. StudLites are a flexible solution literally and figuratively, in that they can be positioned to create whatever size lightbar you might envision. Brickstuff also makes a Universal Lightbar that features a 1× plate with LEDs preinstalled on top, ready to be covered by translucent elements. Universal Lightbars come with a small lighting effect controller board that offers several flashing patterns.

ANIMATING VEHICLE LIGHTS

You can animate LifeLites' 3 mm through-hole LEDs using the LifeLites eLite Advanced lighting effect controller. In fact, from sirens to aircraft navigation lights to rocket engine exhaust, vehicles present many opportunities to incorporate animation effects. For more on animating LEDs, see Chapter 9.

Figure 6-2: Emergency vehicle lightbars

WIRES AND BATTERIES

There typically isn't much space inside LEGO vehicles, especially minifig-scale cars. To make the most of the space available, use LEDs with thin wires and short leads and plan carefully. Start every lighting project by creating a wiring diagram. The diagram in Figure 6-3, for example, is the lighting plan for the truck shown in Figure 6-7. Your wiring diagram should include every LED, expansion board, adapter board, and wire needed, along with any lighting effect controllers. Once all the components are listed, you can determine the space required and locations where everything will be installed. You should also estimate the lengths of wires needed.

Powering your vehicle will also take some planning. If you're installing a vehicle permanently into a larger display, run a power cable out of the bottom and run it through the display to a power supply. If you want to keep the vehicle portable, however, you'll need to make room somewhere in your design for a battery pack. Eight-wide firetrucks, trains, and airplanes can probably fit a standard AAA battery pack, offered by LEGO or third-party manufacturers. Smaller 5-wide and 4-wide vehicles will need something more compact. Several third-party lighting manufacturers make small coin-cell battery packs. The smallest are the size of a 2×2 brick.

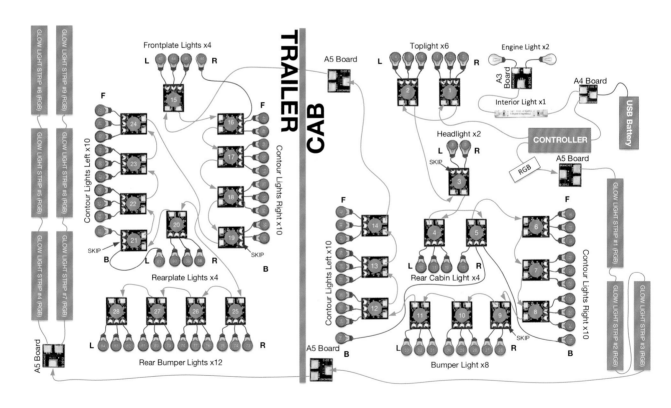

Figure 6-3: Wiring diagram (Brickstuff)

If your vehicle will be powered externally, don't let the power wire hang loose. If it catches on something, it may rip out or damage other wiring. To avoid this, attach any external power wires to a connector board firmly anchored to the vehicle. This way, if the wire gets tugged it will detach from the connector. Alternatively, wrap a power wire around a brick or plate to secure it (BrickLink #2540 bar handle with free ends works well, as shown on the left of Figure 6-4).

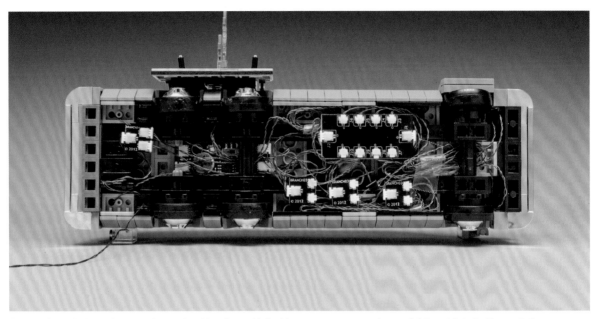

Figure 6-4: Hiding wires on the underside of a vehicle. You can see more views of this vehicle in Figure 6-9.

EXAMPLES OF VEHICLE LIGHTING

With a good understanding of the lighting needs for vehicles, let's now take a look at some great examples of custom-lighted vehicles made by AFOLs. These builds range from small cars to gargantuan spaceships, but they all have lighting techniques that you can learn from.

FORD E-450 AMBULANCE

The ambulance in Figure 6-5 shows the value that flashing lights add to emergency vehicles. The builder used pico LEDs to create headlights, emergency lights, and even the lightbar. They're all animated using a PFx Brick (see Chapter 9 for more on this device). Note the use of custom stickers for added realism.

Figure 6-5: *Ford E-450 Ambulance* (Michael Gale)

Figure 6-6: *Sandcrawler* (Marshal Banana)

SANDCRAWLER

The alien vehicle in Figure 6-6 is as massive
as it is beautiful. The builder spent nearly six
months building this monster, which stands
18 inches tall. It features complete lighting
using only standard LEGO Power Functions
LEDs. Translucent yellow bricks filter some of
the LEDs, giving them a yellow tint.

Figure 6-7: *Peterbilt Quint Axle Dump Truck* (Dennis Glaasker)

DUMP TRUCK

Figure 6-7 shows a 1/16 scale truck. At over 4 feet long, it's large enough to incorporate full navigation and operational lighting. Over 100 Brickstuff LEDs were installed. They run off a USB battery power supply operated by remote control.

Figure 6-8: *Flying Delorean* (author)

FLYING DELOREAN

A lot of custom Delorean MOCs have been built—many before, and some after, LEGO released its official *Back to the Future* set. The MOC in Figure 6-8 is unusual in that it features underside details and full lighting for the flying version. 24 Brickstuff LEDs were installed, along with 20 inches of blue EL glow wire. The green lights underneath are animated using a LifeLites eLite Advanced controller. Three separate voltages drive all the lighting, which creates a lot of wires. These wires all run to the base through a hidden black support rod, just like the miniatures used in filming the movies. Another way that you can create the glowing wire effect is by using a filament LED, which provides brighter light than EL glow wire but may be thicker and shorter.

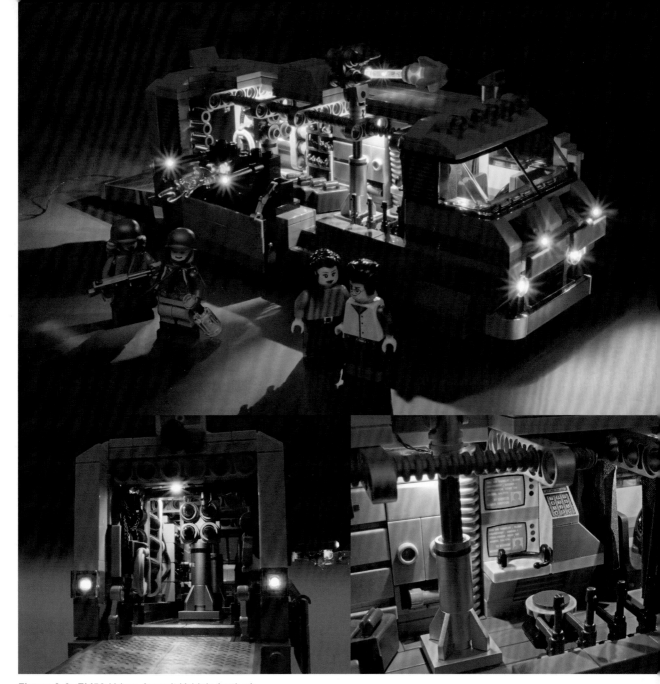

Figure 6-9: *EM50 Urban Assault Vehicle* (author)

EM50 URBAN ASSAULT VEHICLE

Figure 6-9 shows a scale-accurate LEGO version of the customized GMC recreational vehicle from the Bill Murray and Harold Ramis film *Stripes*. Removable panels permit access to weapons systems and full interior details.

14 Brickstuff LEDs provide interior lights, headlights, taillights, and animated weapons systems. As you saw in Figure 6-4, the underside is crammed full of wires and a lighting effect controller.

MILLENNIUM FALCON

There have been many well-built minifig-scale *Millennium Falcon*s over the years. Figure 6-10 shows one of the best. It features screen-accurate details and full lighting inside and out. The blue glow of the engines is made using Brickstuff LEDs backlighting a curved wall of trans-clear 1× plates.

Figure 6-10: *Millennium Falcon* (Marshal Banana)

STAR DESTROYER CHIMAERA

At over 6 feet long, the Star Destroyer in Figure 6-11 is the largest made in LEGO to date. With tons of greebling (complex details added to the exterior) and extensive use of LED under-cabinet lighting, it appears eerily like the scale filming miniatures used in the *Star Wars* movies. LEDs typically run relatively cool, but those used in this ship are so bright and numerous that the operating temperature inside the model is 104 degrees Fahrenheit. The ship requires special vent holes in the top and bottom to dissipate the heat by convection!

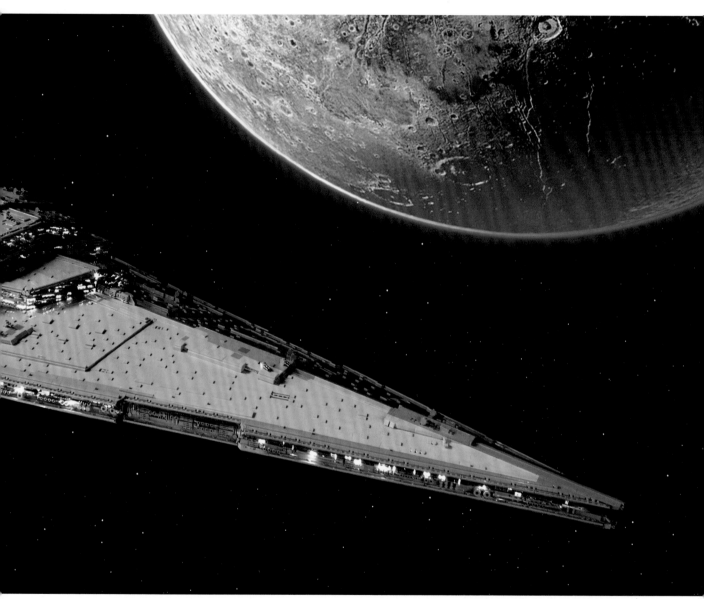

Figure 6-11: *Imperial Star Destroyer Chimaera* (Jerac)

Figure 6-12: Project 6: Mini monster truck

PROJECT 6: BUILD A MINI MONSTER TRUCK

Now it's your turn to experiment with vehicle lighting by lighting your own mini monster truck (see Figure 6-12). This build utilizes six pico LEDs (four warm white and two red). You'll get practice running the wires between plates and bricks and then gathering all the wires to run to the battery pack in the center.

The cab of this truck contains a 2×2 stud opening, three bricks high, where you can fit a small battery pack. Alternately, you can place a small distribution board in this space, connect all the lights to it, and then run a single wire out of the bottom of the truck to a larger battery pack or other power supply.

Requirements

- Four warm white pico LEDs, available from Brickstuff or equivalent third-party lighting manufacturers
- Two red pico LEDs
- One power supply and associated connecting wires and distribution board, as needed

2x
3023
Black

4x
3713
Light Bluish Gray

2x
36840
Light Bluish Gray

2x
99781
Black

2x
48336
Light Bluish Gray

1x
3022
Black

2x
3069bpb0275
White

2x
2817
Black

8x
85861
Yellow

3x
3660
Black

4x
6580c01
White

2x
3710
Yellow

4x
18651
Black

2x
3023
Red

2x
3005
Blue

1x
28802
Black

2x
3023
Orange

2x
4070
Blue

1x
29120
Blue

1x
15068
Blue

1x
4865b
Trans-Black

1x
29119
Blue

2x
15573
Blue

1x
3710
Blue

1x
2437
Trans-Black

10x
3023
Blue

1x
2431
Blue

2x
4073
Trans-Clear

1x
4865b
Blue

2x
30413
Blue

2x
4073
Trans-Red

1x
3004
Blue

1x
28802
Blue

2x
4073
Trans-Yellow

2x
2420
Blue

1x
87079
Blue

1x
2412b
Metallic Silver

1x
3022
Blue

4x
35789
Blue

4x
3069b
Metallic Silver

1

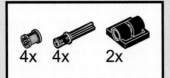

4x 4x 2x

2

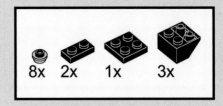

8x 2x 1x 3x

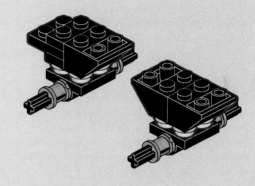

3

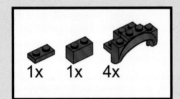

1x 1x 4x

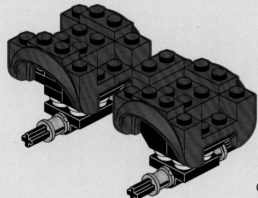

4

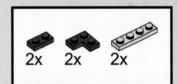

2x 2x 2x

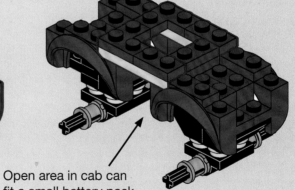

Open area in cab can
fit a small battery pack.

5

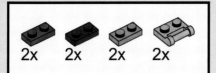

2x 2x 2x 2x

6

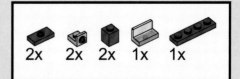

2x 2x 2x 1x 1x

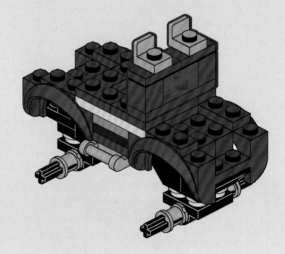

7

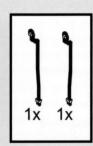

1x 1x

Carefully thread
LED wires between
plates into cab.

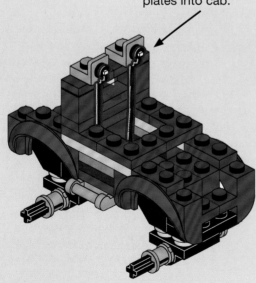

8

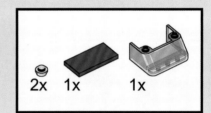

2x 1x 1x

9

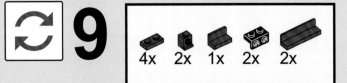

10

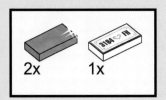

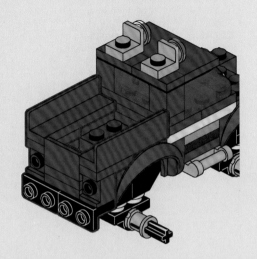

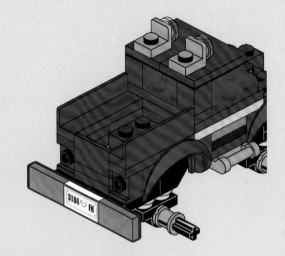

11

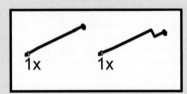

12

Carefully thread
LED wires between
bricks into cab.

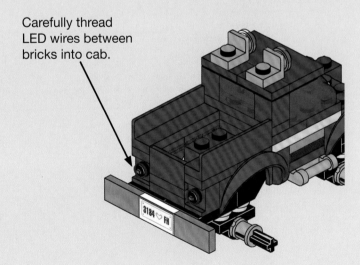

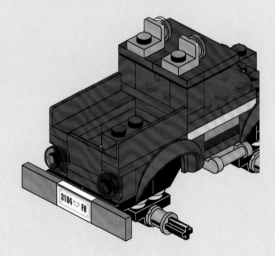

13

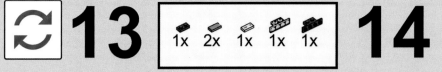
1x 2x 1x 1x 1x

14

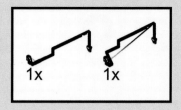
1x 1x

Carefully thread
LED wires between
bricks into cab.

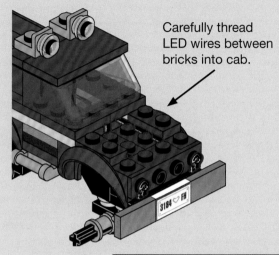

15

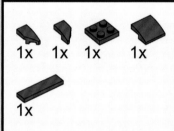
2x 1x

16

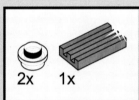
1x 1x 1x 1x

1x

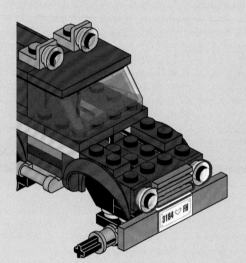

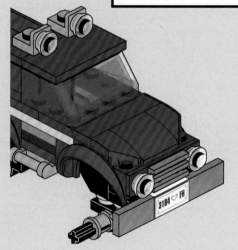

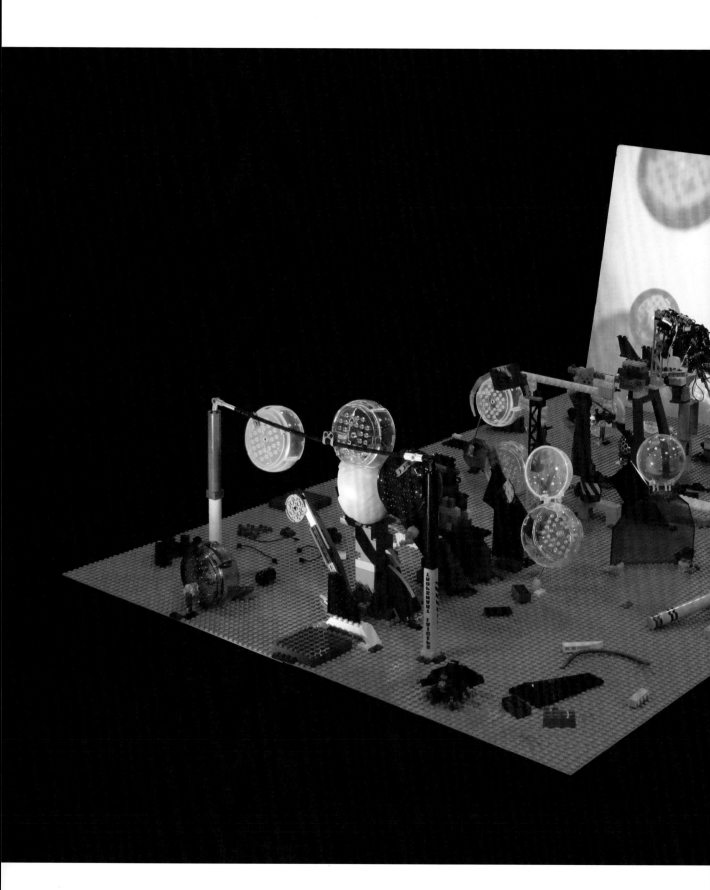

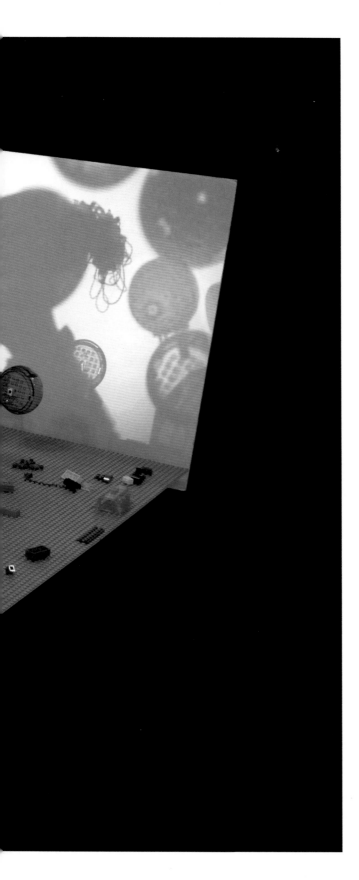

7 SHADOW ART

Shadow is a color as light is, but less brilliant; light and shadow are only the relation of two tones.

—PAUL CÉZANNE, ARTIST

Humans have been casting shadows using their hands since ancient times. Even today, there's something magical about watching the silhouettes of animals and other objects created literally "by hand." It should therefore come as no surprise that contemporary artists have discovered how to use LEGO to delight audiences with the shadows cast by their bricks. Seemingly random arrangements of bricks, when illuminated from just the right angle, suddenly project lifelike silhouettes of familiar objects.

Imagine (Amanda Feuk). A light source out of frame to the left shines on an assortment of LEGO elements, casting a shadow of a young girl.

SHADOW PUPPETRY IN LEGO SETS

LEGO first paid homage to the ancient art of shadow puppetry by including a shadow puppet theater in the 2015 Ninjago set #70751 Temple of Airjitzu, shown in Figure 7-1.

In 2020, LEGO offered a set completely devoted to shadow puppetry, #11009 Classic Bricks and Lights, shown in Figure 7-2. This set allows builders to create a variety of shadow puppets entirely from LEGO elements.

Figure 7-1: LEGO Ninjago set #70751 Temple of Airjitzu contains a miniature shadow puppet theater.

Figure 7-2: LEGO set #11009 Classic Bricks and Lights is devoted to building shadow puppets.

Both sets rely on the principle of shining light on various LEGO elements in order to cast shadows on a viewing surface. The techniques shown in this chapter are based on the same principle, executed on a larger scale.

SHADOWGRAPHY

Shadowgraphy is the art of telling stories through the shadows of hand puppets. It became popular among European magicians in the 19th century and was a precursor to moving picture shows.

MAKING YOUR OWN SHADOW ART

There are two basic approaches to creating shadow art with LEGO: *sidelighting* and *backlighting*. A sidelit creation involves laying LEGO bricks out on a table and casting a shadow onto the table using a light source placed to the side. In this case, only the top edges of the bricks create a thin shadow that can be viewed from above (flip to Figure 7-5 for an example). A backlit creation, by contrast, places bricks between the light source and a screen such that the bricks are surrounded by the stream of light (as in the other examples throughout this chapter). In this case, every side of the bricks can potentially generate shadows. The bricks can be supported by thin bars, transparent panels, or structures hidden in the shadows.

Whichever approach you take, creating your own LEGO shadow art involves three basic steps:

1. Choose an appropriate light source.

2. Find the right bricks to produce the shapes you need.

3. Position the bricks in the way that most effectively produces the shadow you want to create.

Let's go through each step in turn.

CHOOSING A LIGHT SOURCE

To produce hard-edged, well-defined shadows, your light source should be as small as possible. This kind of light is called a *point light source*. For example, the LEGO 2×3 light brick (BrickLink #54930c02) contains a discrete LED that casts sharp shadows because the actual area emitting light is smaller than the head of a pin. Brighter, more powerful LEDs can be found in flashlights. The light source need not be housed in LEGO, although you may find it helpful to make a LEGO-based fixture to hold it in the right location. Some people also build brick enclosures to hide the light source so that it looks like part of the build.

It's possible to use larger lights for your shadow art, but the resulting shadows may be softer and less focused. Candle and lamp flames were used for centuries, for example. Sunlight works, too, because the sun is located so far from the earth.

FINDING THE RIGHT ELEMENTS

Creating shadow art requires a new way of looking at LEGO bricks. Instead of focusing on the brick's color and shape, consider the negative space (the empty space around the shape) the brick creates when casting a shadow. Play with bricks of different contours and visualize how you can combine them to form the different parts of the shadow you want to create. LEGO makes a lot of round elements of varying diameters useful for outside curves: round plates, pulleys, and cylinders, to name a few. Inside curves have fewer options, but some include arches and the 4×4 curved tile. Figure 7-3 shows several elements you can use for creating curves and other shapes. It's also possible to create unique shadows using pieces or clumps of LEGO string. And don't forget that you can approximate complex curves by placing several bricks at just the right angles.

USING TRANSLUCENT ELEMENTS

Although you'll mostly use opaque elements in your shadow sculptures, you can also incorporate translucent elements to add shading and color. Some LEGO plastic elements (cut from translucent plastic sheets) can also cast translucent shadows. Consider varying the placement of translucent elements: the closer they are to the screen, the sharper the shadow edges will be, while placing a translucent element closer to the light source will diffuse the color more.

SETTING UP THE BUILD

The arrangement of the LEGO elements can be several feet deep to fit all the shapes you need to refine a shadow. The space you have is proportional to the sharpness and

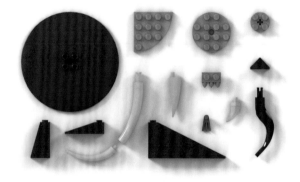

Figure 7-3: LEGO elements well suited to creating shadow art

brightness of your light source. For a sidelit creation, first place a few loose bricks in the center of a table, then position the light at an angle to cast long shadows from the bricks. Next, carefully move and rotate each brick into position to create the edge of the shadow you have in mind.

For backlit creations, you start by positioning the light source and screen. In between, place a couple of baseplates on a table. Shadows that extend offscreen can hide the supports for the bricks (flip to Figure 7-6 for an example). If a shadow is completely surrounded by the light (for example, the shadow of a word), then the bricks will need to be supported by

thin bars, Technic axles, or trans-clear panels. Positioning each brick is a matter of trial and error . . . and patience.

In planning a shadow, it helps to think in terms of stacked planes. For example, the outline of a rabbit is really a collection of curves and other shapes, as shown in Figure 7-4. These shapes can be conveniently separated into groups of similar elements: round plates + wedge plates + regular plates + miscellaneous elements. Place similar elements into "planes" between a light source and a screen in any order, and when a light shines through them, the image of the rabbit will appear.

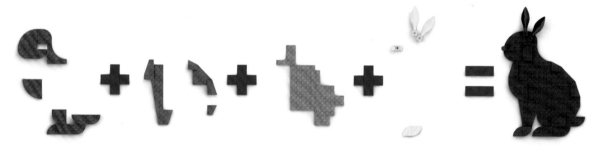

Figure 7-4: Visualize shadow art as a series of stacked planes

EXAMPLES OF SHADOW ART

Now that you understand the basics of how shadow artwork is constructed, let's take a look at some clever examples, all made from LEGO bricks. After studying them, try incorporating some of these techniques into your own MOCs.

FACE

Part of creating shadow art is finding an idea that can be easily recognized when rendered as a shadow outline, or silhouette. Human faces are so recognizable that they've inspired a popular genre of silhouettes called *cameos*.

The sidelit example in Figure 7-5 is rendered using only 12 carefully placed basic bricks and plates on a table. When light is cast from an angle, this seemingly random arrangement of bricks casts the shadow of a human face.

Figure 7-5: *Face* (author)

IN A GOOD BOOK

The build in Figures 7-6 and 7-7 uses hundreds of bricks to create a menacing dragon. But there's more here to see. Inside the main dragon shadow is a carefully crafted blue knight raising his green sword. And note the colorful flames that gently frame this scene. The flame elements are placed close to the light to render their shadows soft and diffused.

Figure 7-6: *In a Good Book* (Amanda Feuk)

Figure 7-7: *In a Good Book*, close-up detail

Figure 7-8: *Magic Angle Sculpture* (John V. Muntean)

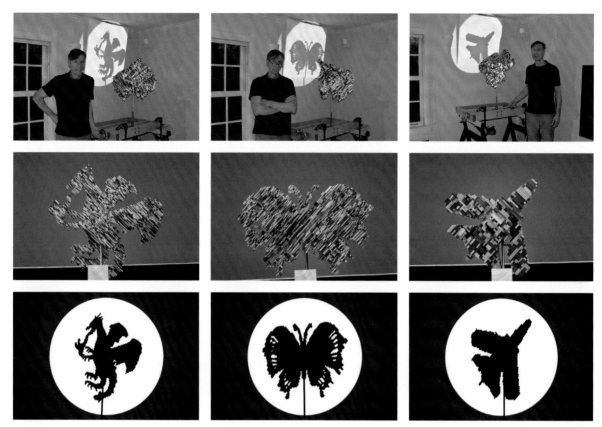

Figure 7-9: *Magic Angle Sculpture*, various angles

MAGIC ANGLE SCULPTURE

Another shadow art technique uses sculptures that cast different recognizable shadows when lit from various angles. Figure 7-8 shows one amazing example.

The build, which rests on a pole, looks like a jumbled collection of LEGO bricks, but when lit from just the right angle and slowly rotated, it reveals the shadow of a dragon, a butterfly, and a jet airplane (see Figure 7-9). The artist's website (*https://jvmuntean.com*) features a video showing how this build appears as it rotates.

PROJECT 7: BUILD A SHADOW SCULPTURE

Now it's your turn to experiment with shadow art by building your own shadow sculpture (see Figure 7-10). This build utilizes a LEGO 2×3 light brick (BrickLink #54930c02) to project an image from a seemingly random collection of parts onto a white panel. Build the sculpture as shown. When finished, press the button on the light brick to reveal the shadow image. For best results, view the shadow sculpture in a dark room. See anything magical about it?

Requirements

- One LEGO 2×3 light brick (BrickLink #54930c02) or equivalent

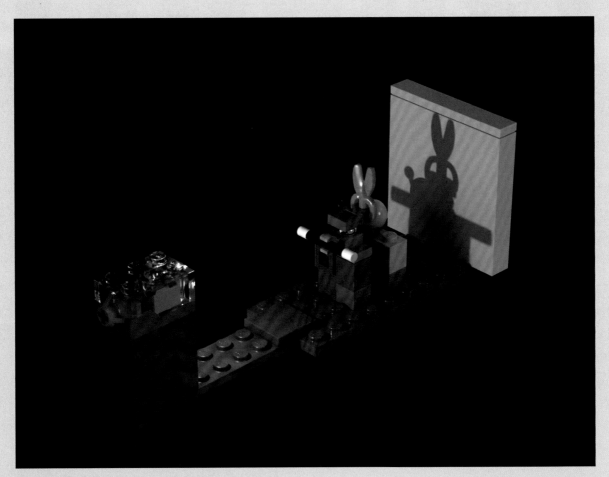

Figure 7-10: Project 7: Shadow sculpture

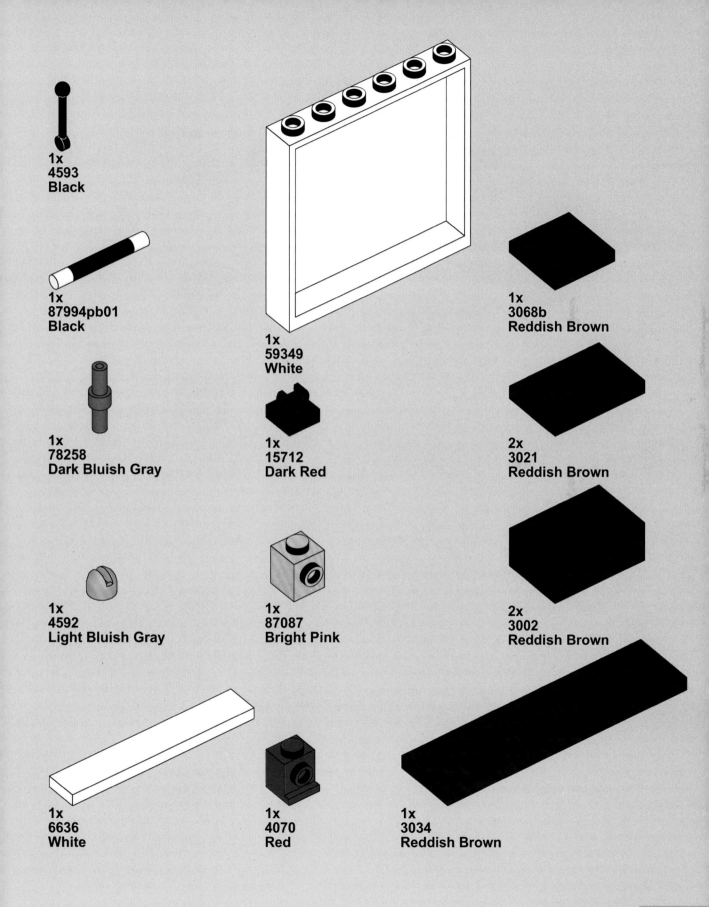

1x
4593
Black

1x
87994pb01
Black

1x
78258
Dark Bluish Gray

1x
4592
Light Bluish Gray

1x
59349
White

1x
15712
Dark Red

1x
87087
Bright Pink

1x
3068b
Reddish Brown

2x
3021
Reddish Brown

2x
3002
Reddish Brown

1x
6636
White

1x
4070
Red

1x
3034
Reddish Brown

1x
15573
Dark Azure

4x
3024
Blue

1x
3030
Reddish Brown

1x
4085d
Blue

1x
61252
Orange

2x
3005
Dark Purple

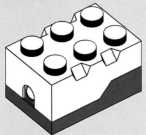

1x
54930c02
Dark Bluish Gray

1x
11010
Chrome Gold

1x
32474
Fabuland Pastel Green

1x
18920
Flat Silver

1

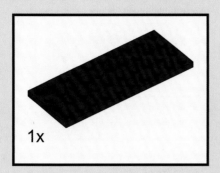

1x

2

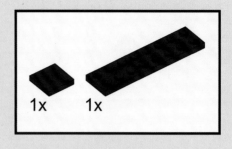

1x   1x

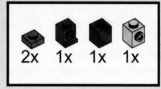

3

2x   1x

4

2x   1x   1x   1x

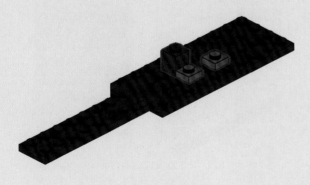

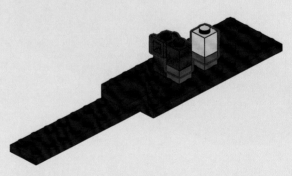

5

1x 1x 1x 1x

6

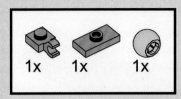

1x 1x 1x

7

1x 1x 1x 1x

8

1x 1x

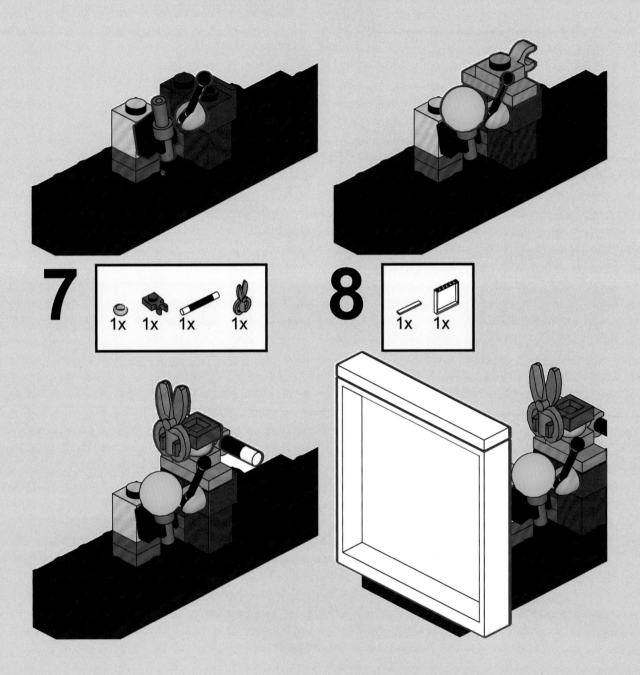

9

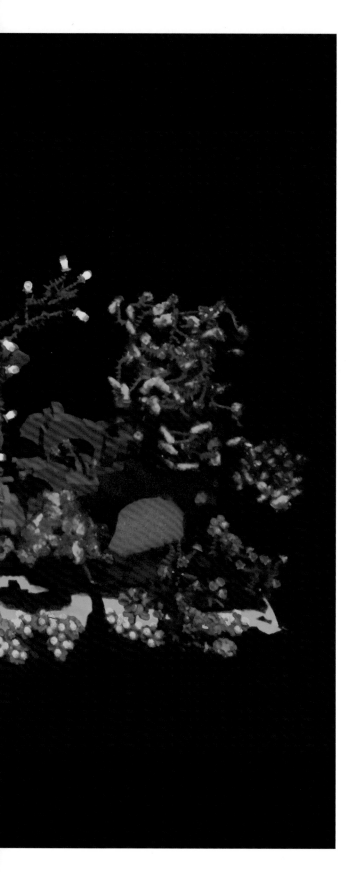

8 GLOWING BRICKS

The fascinating thing is not to show the source of light, but the effect of light.

—EDGAR DEGAS, ARTIST

This chapter, perhaps more than any other in this book, explores the creative possibilities of using lit bricks as art. Some of these MOCs use light to reinterpret real-world objects in abstract, unfamiliar ways. This heightens their emotional impact and conveys a pure fascination with light itself. Like the stained glass windows in the cathedrals of old, MOCs made with glowing bricks can radiate a rainbow of interplaying colors that enchant our senses.

A River Runs Through It (Barbara Hoel). This LEGO sculpture uses special bricks that glow under fluorescent light (blacklight).

Figure 8-1: *Hope Castle* (Reed Yaeger). This 7-foot-wide castle is made entirely of transparent clear bricks. Color-changing LED strips run throughout the castle, creating different color schemes.

USING GLOWING BRICKS

Three different LEGO materials can produce a glowing effect: translucent bricks, glow-in-the-dark (GITD) bricks, and fluorescent bricks. To create a glowing build, you can use one of these options exclusively or combine them.

Beyond offering a wide range of translucent, GITD, and fluorescent bricks, LEGO is great for making glowing sculptures because it provides a complete framework in the form of bricks, panels, and plates, as well as a huge library of supporting elements. Together, these features enable you to position, flag,

and filter glowing elements in countless creative ways.

TRANSLUCENT BRICKS

LEGO has bricks in over 20 transparent colors that allow light to glow and refract in fascinating ways. You can light these bricks from underneath or behind, using LEDs to either add spots of accent color to MOCs or radiate light outward, projecting it onto other surfaces and bricks. Some builders go to the extreme and create entire MOCs from translucent bricks, as in Figure 8-1. In this case, strips of commercial LEDs are typically embedded in the base, face up, to light the large surface areas.

When building large translucent MOCs, be careful to plan how light colors will radiate through the structure. All translucent bricks act like subtractive color filters. For example, if white light is projected through translucent red bricks, the red bricks will pass only red light and block other wavelengths. This means that if you have blue translucent bricks on the other side of the red bricks, the blue bricks will appear black, since no blue light is being allowed through. However, filtering white light through magenta-colored bricks will allow some blue and red light to pass through, so the blue bricks on the other side will appear blue.

WHERE TO FIND TRANSLUCENT BRICKS

BrickLink offers a table of past and present LEGO colors, allowing you to see all elements ever made in any given translucent color. If you envision lighting translucent elements, start with this table and examine all the elements that are available in the colors you want for your MOC.

GLOW-IN-THE-DARK BRICKS

LEGO bricks that glow in the dark are a favorite for adding interest to nighttime MOCs (see Figure 8-2). When exposed to room light or another light source, GITD elements are "energized" and will then glow in a darkened room.

These elements can permit a contrasting, alternate perspective on a MOC when viewed in the dark versus in the light. While GITD elements aren't as bright as LEDs and don't glow for very long, they're easy to build with and don't require any electrical wires.

GLOW-IN-THE-DARK HISTORY

The first proper LEGO system element to be molded in GITD plastic was the Ghost (#2888), which appeared in the 1990s. Since then, over 80 different GITD elements have been produced.

Figure 8-2: *Opee Sea Killer* (Matt De Lanoy). The glow-in-the-dark elements heighten the eeriness of the subject.

FLUORESCENT BRICKS

Some LEGO bricks fluoresce under ultraviolet light as a result of the type of plastic used. Ever since AFOLs discovered this odd quirk, they've exploited it to build amazing creations that magically glow under blacklights (for an example, see the opening image of this chapter). Artists typically use commercial UV lamps to light their builds, but note that Brickstuff makes a UV LED specifically for this purpose as well.

Not every LEGO brick fluoresces. Not even every brick of the same color can be counted on to fluoresce—because plastic formulations change over time, some bricks of a certain color might fluoresce, while later runs of these same bricks won't. You can refer to artists' works to see what colors they've used, but your best bet is to get a blacklight and start experimenting with the bricks you have on hand.

WARNING Be sure not to stare at UV LEDs. The UV radiation can be very strong and damage your eyes, even when the LED doesn't look very bright.

EXAMPLES OF GLOWING BRICKS

With that introduction to the three different glowing effects, let's now take a look at how brick artists have used them to create impressive pieces of art from LEGO bricks. Study these examples and then try incorporating some of the techniques into your own MOCs.

ICE SCULPTURES

The life-size sculptures in Figure 8-3 are made from about 30,000 trans-clear 1×2 bricks lit from the inside with LED strips. Note the bright contours and patterns caused by the refraction of the light interacting with the bricks. The large scale of the sculptures manipulates light in unique ways.

Figure 8-3: *Ice Sculptures* (Chairudo)

TOXIC LAKE

The flat panel shown in Figure 8-4 uses an LED artist's light panel underneath a trans-clear baseplate that has been covered with various shades of cheese slope bricks. The direction of each slope controls the way light refracts through the brick and can create some amazing patterns—in this case, waves. Because genuine LEGO trans-clear baseplates are quite expensive at this time, brick artists you can often turn to clone baseplates for projects like this. You can use this same technique with red, orange, and yellow slopes to create rivers of lava, or you can combine different trans-colors to create a mosaic stained glass window.

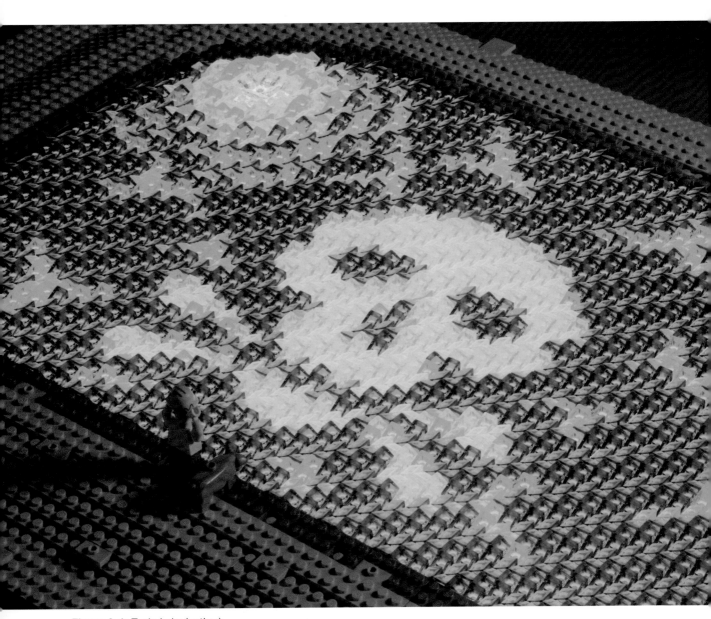

Figure 8-4: *Toxic Lake* (author)

Figure 8-5: *Stained Glass Windows* (Alyska Bailey Peterson)

STAINED GLASS WINDOWS

The stained glass windows in Figure 8-5 were created using a *studs-up* technique: the trans-colored bricks are stacked vertically, with their studs pointing up, building up the resulting images row by row. This contrasts with the cheese slope technique shown in Figure 8-4 and the layered trans-color technique used in Figure 8-10, both of which are *studs-out*. Studs-out building lets you create different looks with just a few different trans-colors and produces a strong mosaic, but the technique requires costly trans-colored plates or baseplates. Meanwhile, studs-up building doesn't require rare baseplates, but the resulting mosaics are very delicate at sizes larger than a few inches. It's best to transport larger studs-up mosaics by building a strong frame around them to hold everything together, as in the stained glass windows shown here.

TIFFANY-STYLE STAINED GLASS LAMP

LEGO cheese slopes are frequently used to make LEGO stained glass MOCs because of their unique geometry and availability in so many translucent colors. The build in Figure 8-6 combines cheese slopes and other trans-colored elements to render fireflies in a Tiffany-style miniature table lamp. The elements are fitted into trans-clear 1×6×5 panels and backlit using a bright 1 watt LED. The pearl gold–colored tassels, which pick up and diffuse the light coming from below the shade, add interest.

Figure 8-6: *Tiffany-Style Stained Glass Lamp* (author)

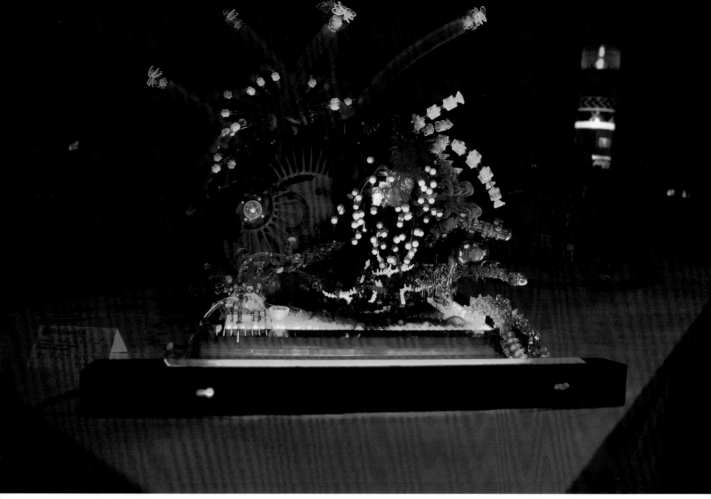

Figure 8-7: *Jenn's Fish* aka *My First MOC Evah!!* (Jenn Wagner)

JENN'S FISH

The scene shown in Figure 8-7 depicts glowing life-forms in the dark depths of the ocean. The artist combines an array of fluorescent LEGO elements in an arrangement specifically meant to be lit by a blacklight tube placed in front of and below the model. This creates an eerie glow of blues and purples, with contrasting reds and greens that pop. Note this MOC uses a variety of elements, including hoses and tubes alongside the more common hard-plastic bricks.

GHOST SHIP

In Figure 8-8, the artist incorporates a variety of GITD LEGO elements. Flat tiles, modified 1×1 teeth, and dinosaur tails all highlight the ship's lines and curves. Glowing seaweed adds to the nautical theme. The builder has even coated the sails with GITD paint, which you can find at your local hobby store. All this combines to create a very mysterious ghost ship . . . when viewed in the dark, of course.

Figure 8-8: *Ghost Ship* (Jens Ådne J. Rydland)

CRYSTALLINE TREE

The tree in Figure 8-9 is built almost entirely from transparent purple Hero Factory ball joints, chains, and Bionicle Spine Armor elements. These are all wound around an armature built from black Technic axles and angled pin connectors. Strings of wired LEDs are intertwined with the brick elements to create a delicate yet stunning look when lit. This MOC won a lighting award at BrickFair Virginia.

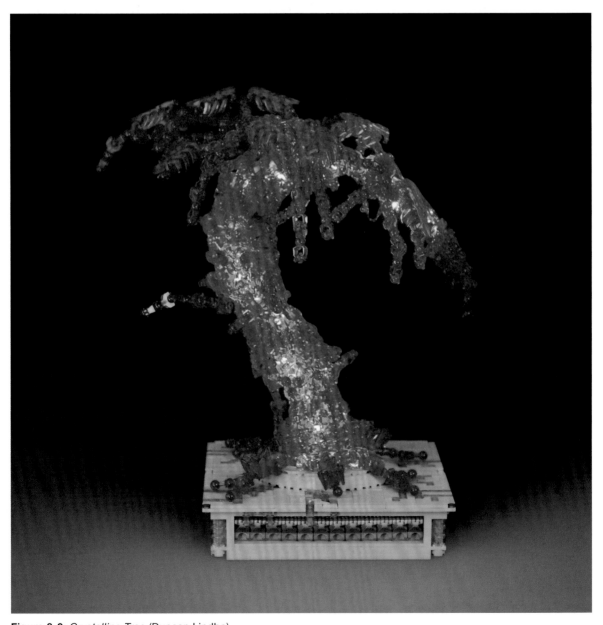

Figure 8-9: *Crystalline Tree* (Duncan Lindbo)

LINCOLN PORTRAIT

The MOC in Figure 8-10 is a LEGO implementation of a popular image of Abraham Lincoln first published in a 1973 *Scientific American* article titled "The Recognition of Faces," by Leon Harmon, a researcher at Bell Labs. It uses the bare minimum number of pixels needed to make this familiar face recognizable. LEGO trans-colored bricks are frequently overlapped to provide deeper color saturation or new colors. In this case, trans-black plates, stacked at various depths and positioned studs out, are backlit by an LED light box to produce up to seven shades of gray (the image on the right); the more plates in the stack, the darker the shading of that "pixel." Viewed without the backlight on, the same MOC produces only three discernable levels of shading (image on the left).

Figure 8-10: *Lincoln Portrait* (author)

PROJECT 8: BUILD A SUN CATCHER

Now it's your turn to experiment with glowing bricks by building your own sun catcher (see Figure 8-11). This is the only project in this book that utilizes the sun itself for light . . . no LED required. Carefully place the small plates and cheese slopes into the 1×6×5 panel "frame" and then hang it in a window to brighten your day.

Requirements

- One sun. Look outside . . . just not directly at it.

Figure 8-11: Project 8: Sun catcher

1x
2654
Black

1x
74698
Black

1x
3070b
Trans-Black

1x
3024
Trans-Black

5x
54200
Trans-Black

2x
3069b
Trans-Black

1x
3024
Trans-Clear

12x
3023
Trans-Clear

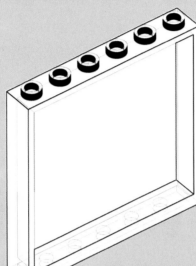

2x
59349
Trans-Clear

6x
54200
Trans-Orange

1x
3024
Trans-Neon Green

5x
54200
Trans-Neon Green

2x
3024
Trans-Green

5x
54200
Trans-Green

1x
3024
Trans-Dark Blue

29x
54200
Trans-Dark Blue

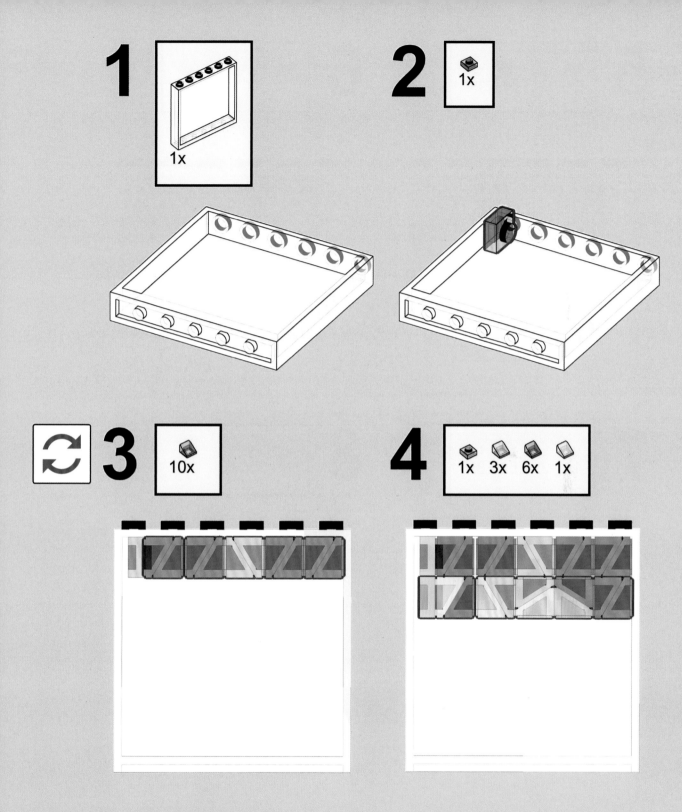

5

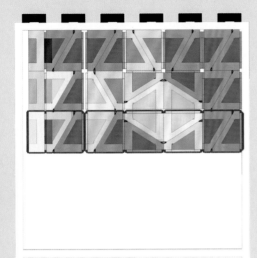

6

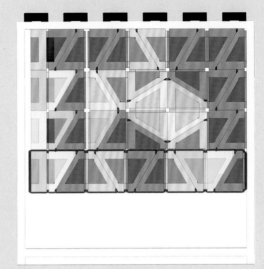

7

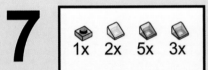

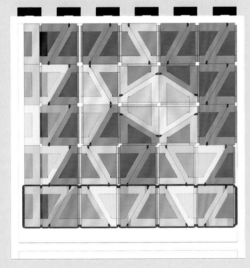

8

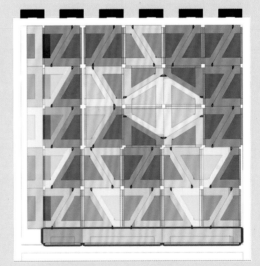

1x 12x

10

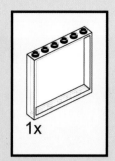

1x

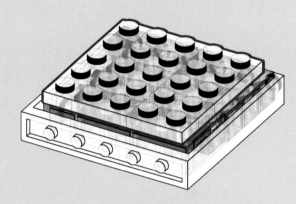

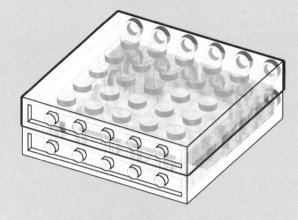

11

1x 1x

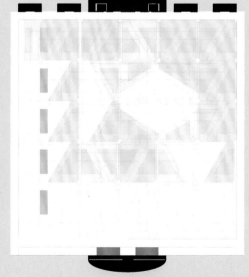

12

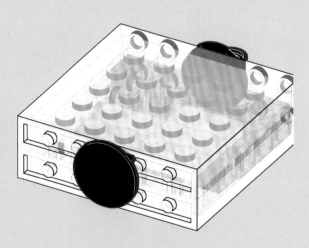

9 DYNAMIC LIGHTING

What is soul? It's like electricity—we don't really know what it is, but it's a force that can light a room.

—RAY CHARLES, MUSICIAN

There are times when static lighting just can't capture the essence of a scene. What would a dank dungeon be without the flicker of a torch, the Las Vegas Strip without animated neon signs, or a rock concert without the sparkle of a thousand camera flashes? From emergency vehicles to theater marquees to traffic signals, flashing lights are everywhere. Even our futuristic visions of outer space are dominated by blinking control panels, pulsating engine lights, and strobing laser beams.

Star Portal (author). This shadow box diorama incorporates three dynamic lights: (1) a tablet computer in back, projecting a custom-made animation loop of bubbling water; (2) an LEC above, casting fluctuating shadows behind the minifigures; and (3) the yellow rotating beacon on the left wall.

Now that you've learned about the art of incorporating static lights into your LEGO creations, you're ready to take the next step and bring your MOCs to life through *dynamic lighting*. In this chapter, we'll discuss ways to animate lights using *lighting effect controllers* *(LECs)*, small electronic circuit boards that control one or more LEDs using various patterns. Some LECs will make lights strobe, flash, or flicker at random. Others can make a number of LEDs light up in specific patterns, like the marquee on a theater.

LIGHTING EFFECT CONTROLLERS

The simplest LEC is so tiny that it's molded into the package of a discrete LED. This is how premade flashing LEDs work. The circuit is built into the LED itself and requires no additional space or controls. Of course, the rate of flashing is fixed. Some strands of commercial LEDs also come with flashing circuits built into their battery box.

For more control, several third-party companies offer dedicated LECs that allow you to connect LEDs to specific channels, which are controlled by a sequence. The sequence determines when each channel of LEDs is lit. These controllers vary in several key capabilities, including the number of channels supported; the number of LEDs that can be attached to a channel; the number of sequences offered; and the ability to adjust the speed, brightness, and other attributes of the sequences. Some common third-party LECs are shown in Figure 9-1. We'll discuss each one in this section.

LIFELITES ELITE ADVANCED

The LifeLites eLite Advanced controller accommodates eight separate channels and is preloaded with dozens of popular lighting sequences. The whole LEC fits into the space of a 2×4 brick. One very useful sequence is the *lighthouse*, in which each of the eight channels fades up and down in succession. Run at high speeds, this sequence can simulate fires; this works best with a mix of red and yellow LEDs randomly distributed inside a trans-orange structure built to look like flames. Other sequences simulate torches, theater marquees, lasers, rocket engines, and more.

BRICKSTUFF LECS

Brickstuff makes LECs for many use cases, including emergency vehicle lightbars, aircraft navigation lights, torches, welding, and machine gun fire. Its latest BrickPixel™ Lighting Effect Controllers support high-amperage channels that can flicker and

Figure 9-1: LEC examples, clockwise from top left: eLite Advanced (LifeLites), torches (Brickstuff), BrickPixel™ Lighting Effect Controller (Brickstuff), PFx Brick (Fx Bricks), Rotating Sequence Effects Board (Light My Bricks), and Light Genie (MRC)

flash hundreds of LEDs. Some incorporate remote control and synchronized sound, and can even run motors!

FX BRICKS PFX BRICK

Fx Bricks makes the PFx Brick, a powerful lighting effect, sound, and motor controller all in one package. The PFx Brick is specially made for vehicles. Its 10 light channels support about 5 LEDs per channel. The lights can dim, flicker, or be combined for special sequences, including simulated strobe flashers, lasers, spaceship engines, and locomotive lighting patterns. The lighting and sound effects can be triggered and modulated by the motor control outputs, too. By combining light, sound, and motion, you could, for example, synchronize the muzzle flash and sound of a chain gun with the gun's rotation.

LIGHT MY BRICKS ROTATING SEQUENCE EFFECTS BOARD

Light My Bricks makes several LECs, including a Rotating Sequence Effects Board that provides an inexpensive solution to animating LEDs. It controls eight channels arranged in two sets of four that run the same sequence, flashing LEDs in sequence. You can use it to animate theater marquees, laser beams, rocket engines, and fire effects.

MRC LIGHT GENIE

There are other commercial LECs available that are not specifically intended for use with LEGO. One of these is the Light Genie made by Model Rectifier Corporation (MRC). It's designed specifically for model railway and scale miniatures hobbies. The Light Genie has several unique features if you need *lots* of lighting effects. First, it offers 12 channels with over 20 effect sequences. Second, these 12 channels can be organized into 5 separate groups (MRC calls them *zones*), each controlled by a different sequence. This way, you can have multiple sequences operating in parallel, something that would require separate LECs with other systems. Additionally, multiple Light Genies can be operated together and controlled through the same wireless transmitter. This is a very capable system that can be customized to animate everything from airport runway lights to traffic control signals along an entire stretch of road.

DON'T GO OVERBOARD

Even in a large LEGO display, a single dynamic effect can go a long way . . . meaning it's easy to crowd a scene with too many animation effects that can potentially overwhelm the viewer (and strain the builder's wallet). For any given display area, say each 8-foot table, select no more than two or three items to animate with light.

EXAMPLES OF DYNAMIC LIGHTING

With that background on LECs, let's now take a look at how they can be combined in MOCs to create some amazing effects.

AKATOR THRONE ROOM

The challenge with the sci-fi build shown in Figure 9-2 was to create the illusion of a room spinning without actually rotating the whole build. The solution was to motorize only the small pit in the center and then add a series of lights along the back that flash quickly left to right, synchronized with the motor. The lights are driven by a LifeLites eLite Advanced LEC with a marquee pattern, creating a deliberately dizzying effect.

Figure 9-2: *Akator Throne Room* (author)

Figure 9-3: *Snake Pit* (author)

SNAKE PIT

The ancient temple depicted in Figure 9-3
is inhabited by hundreds of snakes! The
only thing keeping them at bay are a dozen
flickering torches, driven by several LECs.
The flickering effect really adds drama to the
flames. Also present, but not shown, is a peri-
odic flash of lighting from above, driven by a
flashing LEC.

Figure 9-4: *Sinking Ship* (author)

SINKING SHIP

The doomed ship in Figure 9-4 is engulfed in an inferno as it sinks beneath the waves. The fire is animated with eight yellow LEDs and a LifeLites eLite Advanced LEC running a lighthouse sequence. This draws the viewer's eye and heightens the drama. You'll find that the lighthouse sequence (also called *chase*, *loop*, or *marquee* in other products) is very flexible and can be used for everything from theater marquees to fire to, well, lighthouses. In addition to the dynamic lighting, there are also cool blue LEDs projected from the right and left to highlight the contour of the hull and the waves.

PROJECT 9: BUILD A THEATER MARQUEE

Now it's your turn to experiment with dynamic lighting by building your own animated theater marquee (see Figure 9-5). The instructions use a Rotating Sequence Effects Board from Light My Bricks, but you can also use an LEC from Brickstuff, LifeLites, or Brick Loot that has a chase (also called marquee or rotating) sequence. Different LECs may support a different number of lights or channels. There's just enough space in the back of the build to conceal a coin-cell battery pack if you wish. Or, for longer runtime, you can run a wire out of the back to a larger battery pack or alternate power supply of your choice.

The text, made from decorated 1×1 LEGO tiles, can be changed to anything you want. Incorporate this animated marquee into any theater or club you may have in your layout. Or put your name on it and wear it as a name tag while visiting your local LEGO fan club event!

Requirements

- Eight warm-white pico LEDs, available from Brickstuff or equivalent third-party lighting manufacturers.
- One Light My Bricks Rotating Sequence Effects Board. Alternatively, you can use an equivalent LEC from Brickstuff, LifeLites, or Brick Loot that has a chase sequence.
- One power supply, plus distribution boards and connecting wires, as needed.

Figure 9-5: Project 9: Theater marquee

1x
3070bpb016
Black

1x
3070bpb023
Black

1x
3070bpb013
Black

2x
3070bpb020
Black

3x
3070bpb015
Black

2x
3070bpb017
Black

1x
3070bpb028
Black

1x
3070bpb022
Black

2x
36841
Black

4x
26604
Black

12x
35480
Black

2x
3069b
Black

1x
3034
Black

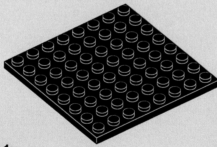

1x
41539
Black

2x
3020
Red

2x
3832
Red

20x
35480
Dark Tan

24x
4073
Trans-Yellow

1

20x

2

12x

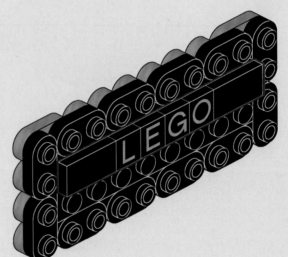

3

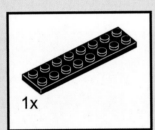
1x

4

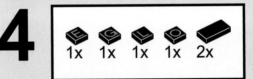
1x 1x 1x 1x 2x

5

2x 1x 2x 1x 1x 1x

   6

4x

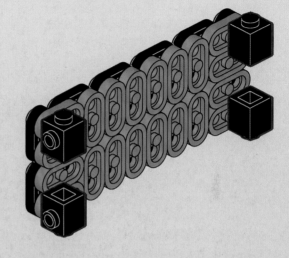

7

2x 2x 2x

8

1x

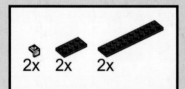

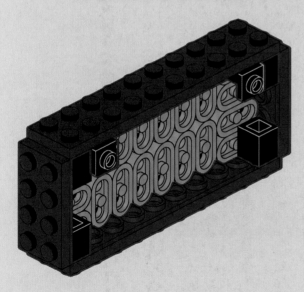

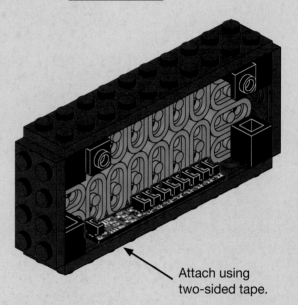

Attach using
two-sided tape.

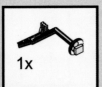

 10

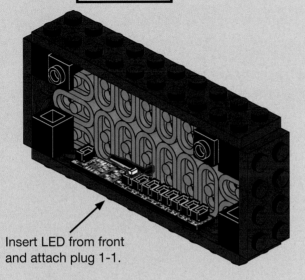

Insert LED from front
and attach plug 1-1.

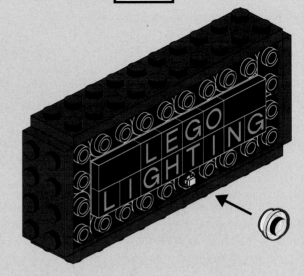

 11

 **12**

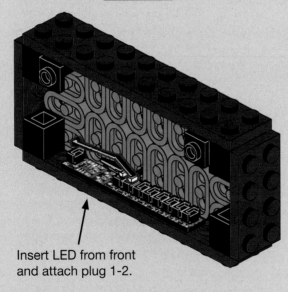

Insert LED from front
and attach plug 1-2.

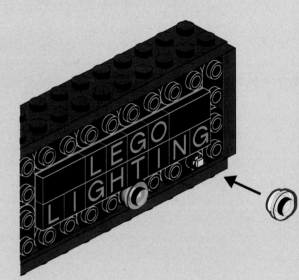

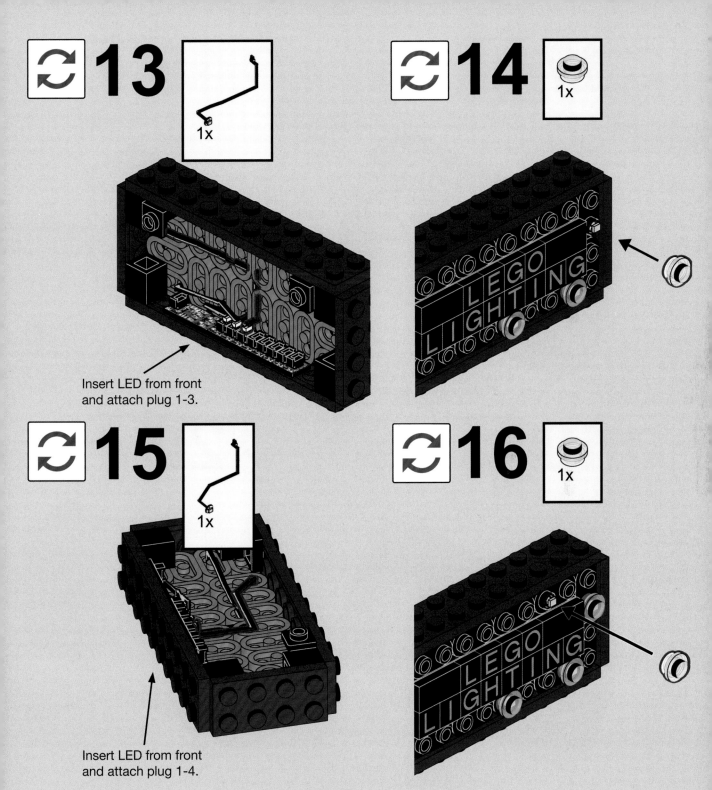

13

1x

Insert LED from front
and attach plug 1-3.

14

1x

15

1x

Insert LED from front
and attach plug 1-4.

16

1x

17

1x

Insert LED from front
and attach plug 2-1.

18

1x

19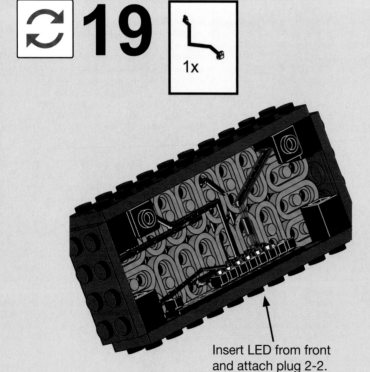

1x

Insert LED from front
and attach plug 2-2.

20

1x

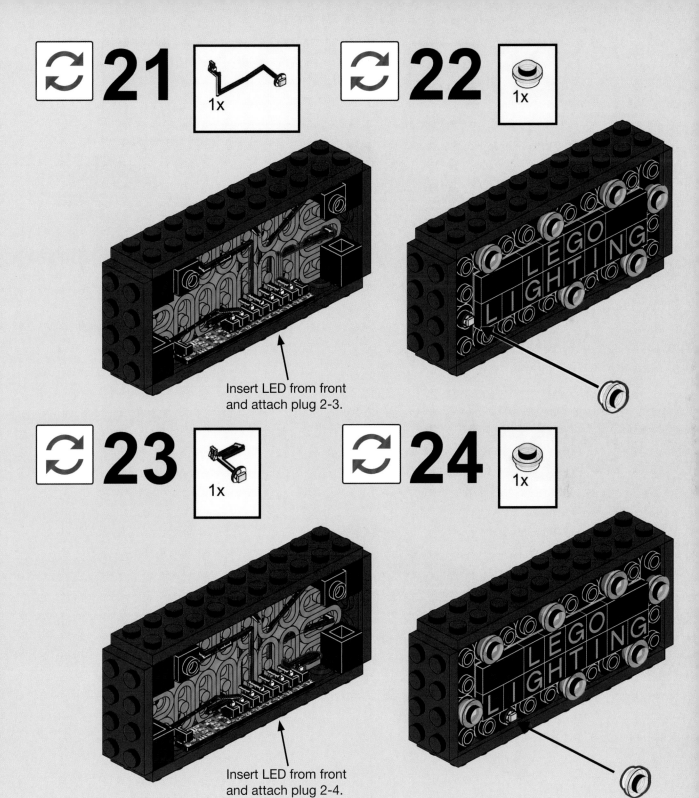

21 1x

22 1x

Insert LED from front
and attach plug 2-3.

23 1x

24 1x

Insert LED from front
and attach plug 2-4.

25

16x

26

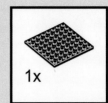

1x

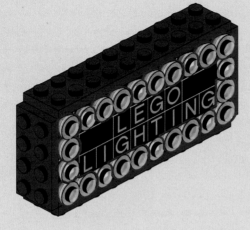

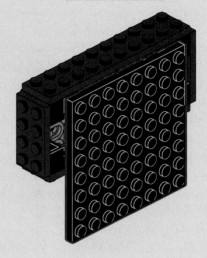

27

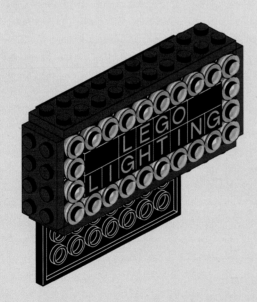

GLOSSARY

ACCENT LIGHT   A light, often artificial, placed to illuminate an architectural detail.

ADAFRUIT   A distributor of LEDs and other electronic components for hobbyist and maker markets (*https://www.adafruit.com*).

AFOL   An acronym for *adult fan of LEGO*, referring to grown-up LEGO enthusiasts.

AMBIENT LIGHT   The sum total of all the light in a space, usually diffused.

ARCHITECTURAL LIGHTING   A field that applies light as a design element to enhance a building's visual aesthetic, its functional usability, and the viewer's experience of the space.

BARN DOORS   Light modifiers for controlling the shape or pattern cast by a light. Typically these are used to keep light off a specific part of a scene.

BASIC LIGHTING   Lighting a scene minimally in order to provide enough light to view it, often consisting of a single lamp.

BATTERY LIGHT   A common term for a type of light made by the LEGO Group that has either a 2×3 or 2×4 stud form factor and contains an LED powered internally by small coin-cell batteries.

BIT LIGHT   A small LED offered by Light My Bricks, equivalent to Pico LEDs sold by Brickstuff.

BRICKFAIR   The name of a series of LEGO fan conventions and public exhibitions held throughout the eastern seaboard of the United States. BrickFair Virginia held its first World of Lights event in 2017.

BRICKFILM   A short movie made by photographing practical brick sets and LEGO characters using stop-motion animation.

BRICKLINK   A retail marketplace for reselling LEGO sets and bricks (*https://www.bricklink.com*). LEGO lighting elements from past years can usually be found for sale here.

BRICK LOOT   A manufacturer of lighting products for the LEGO and hobby markets (*https://www.brickloot.com*).

BRICKSTUFF   A manufacturer of lighting products for the LEGO and hobby markets (*https://www.brickstuff.com*).

BRICKWORLD   The name of a series of LEGO fan conventions and public exhibitions held throughout the Midwestern United States. In 2009, Brickworld Chicago became the first LEGO fan convention to feature a World of Lights event.

CAFÉ CORNER   A popular line of LEGO town sets consisting of one or more multi-story buildings on a single baseplate.

CHIP LEDS   SMD LEDs approximately 3.2 mm wide. Also a term Brickstuff uses to refer to its smallest series of LEDs.

CINEMATIC LIGHTING   Lighting used creatively to convey the mood of a scene, direct the viewer's attention, or help tell a story.

CIRCUIT   The complete path of an electric current, usually including a source of energy (typically a battery or AC adapter), one or more LEDs, and any other devices the current must flow through to return to the energy source.

COB LED   Short for *chip on board*, a type of SMD LED in which multiple diodes are combined into one package, resulting in more light output and longer life in less space.

COLLIMATE   The process of aligning rays of light so they are parallel, as with the beam of a flashlight.

CONVECTION   A process of transferring heat out of a MOC by moving heated air out through-holes in the top of the build and replacing it with cooler air brought in through holes in the bottom.

DISCRETE LED   A type of LED package in which the electronics are encapsulated in a transparent plastic cylinder with a hemispherical light-emitting end that focuses the light. These are sometimes called *through-hole*, *bare*, or *indicator* LEDs. They're typically 3 mm or 5 mm in diameter, but they can come in larger sizes for specialized applications.

DO-IT-YOURSELF (DIY) LIGHTING   The practice of adding lights by assembling the components on your own. This often means soldering together LEDs, resistors, and wires to a power source. Some third-party lighting manufacturers also use the term to refer to the ability to purchase their lighting components individually, without having to buy them in a larger set.

DYNAMIC LIGHTING   Lighting that changes over time—for example, flashing on and off—to add interest, depth, and realism to a build.

FACADE LIGHTING   The intentional placement of exterior lights on a building or structure to highlight architectural lines, details, and other features.

FILL LIGHT   A soft-edged light used to temper the shadows of a key light or as a secondary light source in a scene.

FLUORESCENCE   The emission of light by a substance that has absorbed light or other electromagnetic radiation. Often ultraviolet wavelengths of light are used to cause fluorescent LEGO bricks to radiate green, orange, or other wavelengths of visible light.

GLOW IN THE DARK (GITD)   A phosphorescent quality wherein an object absorbs light energy and then radiates some of it back in the dark.

GREEBLING   High-density cosmetic details added to the surface of an larger object to give the appearance of scale, complexity, or technological capability.

HID   Short for *high-intensity discharge*, a type of arc lamp that produces light by means of electric discharge between two electrodes. The light produced is very bright and cool in appearance.

INCANDESCENT LAMP   A type of lamp in which light is produced by using electrical current to heat a filament until it glows. The light produced by incandescent lamps is usually warmer in appearance than white sunlight.

INDUCTIVE CHARGING   A technique for transferring power to a device through the air, without wires.

INTERIOR LIGHTING   The overall illumination of a space inside a structure. There are four main styles of interior lighting: natural, ambient, accent, and task lighting.

KEY LIGHT   The main and often brightest light in a scene. Key lights typically cast hard-edged shadows.

LANDSCAPE LIGHTING   The use of light to define the contours of tree lines, walking paths, fountains, and other landscaping elements.

LED PACKAGE   A plastic housing that contains one or more light-emitting diodes. Popular LED package types include discrete, SMD, and COB.

LEKO   A type of key light that projects a focused pattern.

LIFELITES   A manufacturer of lighting products for the LEGO and hobby markets (*https://www.lifelites.com*).

LIGHT & SOUND   A LEGO electronics system introduced in 1985 that integrated 9 V lights, sounds, and motors, using electrically conductive plates that could be built into creations.

LIGHTBAR   An enclosed array of lights mounted on top of emergency vehicles for the purpose of alerting the public or traffic to move out of the vehicle's path.

LIGHT-DIFFUSING FILM   A type of semi-opaque neutral-colored plastic, with a frosted white appearance, used to diffuse light. Sometimes these films are used as lighting gels over lamps. ANOTHER common use is to place them behind windows in miniature buildings to diffuse the light in the miniature room while obscuring the interior details.

LIGHT-EMITTING DIODE (LED)   A semiconductor component that emits light when current is passed through in one direction. LEDs emit a wide variety of colors based on the chemistry of the semiconductor materials used.

LIGHTING DIAGRAM   A drawing that includes every LED, expansion board, adapter board, and lighting effect controllers needed, all connected together with the necessary wires, forming a circuit.

LIGHTING EFFECT CONTROLLER (LEC)   A small electronic circuit board that causes one or more LEDs to animate or flash in one or more patterns.

LIGHTING GEL   A light modifier consisting of a piece of transparent colored plastic or glass used to change a light's color.

LIGHT MODIFIER   Materials and devices used to modify the appearance of light. A lighting gel, for example, can be used to make the light a different color.

LIGHT MY BRICKS   A manufacturer of lighting products for the LEGO and hobby markets (*https://www.lightmybricks.com*).

MARQUEE   A canopy projecting over the entrance to a theater, hotel, or other building that often uses flashing lights to display the name of the venue or attractions therein.

MOC   An acronym for *My Own Creation*, referring to an original design constructed of LEGO bricks, as opposed to a design featured in a commercially available LEGO set. Pronounced "mock."

NANO LEDS   SMD LEDs approximately 1.6 mm wide. Also a term used by LifeLites to refer to its smallest lights and by Woodland Scenics to refer to its smallest Just Plug lights.

NATURAL LIGHT   The light from the sun and sky, which can illuminate an interior space through skylights and windows.

PEERON   A fan-created database of LEGO products and constituent elements (*http://www.peeron.com*).

PICO LEDS   SMD LEDs approximately 1 mm wide. Also a term used by Brickstuff to refer to its smallest LEDs, which consist of a pico-size LED mounted on a board small enough to fit under a 1×1 round LEGO plate. This book uses *Pico* specifically to refer to the Brickstuff product and *pico* generically to refer to any manufacturer's LEDs that meet this form factor.

POINT LIGHT SOURCE   A very small source of light, approximating a point in space.

PRACTICAL LIGHT   An object visible within a scene that emits its own light—for example, a fire, torch, television, computer monitor, or headlights on a car.

PULSE WIDTH MODULATION (PWM)   A method of controlling electrical power intensity by varying the percentage of each cycle that a constant voltage is applied. For example, the LEGO Power Functions system controls a motor's speed by varying the percentage of each cycle (which lasts 1/160 of a second) that a 9 V current is applied. Low power would be only a small fraction of the cycle, half power would correspond to a 50 percent cycle, and full power would apply the voltage 100 percent of the time.

RGB LED   A type of LED that contains separate red, green, and blue LEDs in a single package. The individual LEDs can be set to different brightness levels, combining to produce a full range of colors.

SHADOW BOX DIORAMA   A miniature scene contained in a box viewed through one side in which the viewing angle and lighting are carefully controlled to maximize the storytelling value.

SHADOW SCULPTURE   An arrangement of brick elements that creates a recognizable image when light is shined on it.

SMD LED   Short for *surface-mount device*, a type of LED with a flat package that can be soldered directly to a circuit board.

SPOTLIGHT   A bright key light that throws a long, narrowly confined, hard-edged light.

STATIC LIGHTING   Lighting that illuminates but does not animate or flash.

STRIP LIGHT   An adhesive-backed flexible strip containing multiple LEDs, sold by several third-party lighting manufacturers.

TASK LIGHT   A light, often artificial, placed to illuminate an area where work is done.

TECHNIC BRICK   A type of LEGO brick with 5 mm holes in its side into which 5 mm through-hole LEDs can be inserted. Unlike Technic liftarms, Technic bricks have studs on the top like a regular brick.

TECHNIC LIFTARMS   A type of LEGO element with 5 mm holes in the side into which 5 mm through-hole LEDs can be inserted.

THROUGH-HOLE LED   An LED that is encased in a clear plastic housing with the leads sticking out of the bottom. It is designed to fit through a hole in a part like a panel or LEGO brick.

TRANS-CLEAR   A term in LEGO brick color taxonomy describing a transparent element with no color tint. Older trans-clear bricks can, however, take on a yellow tint over time. For more information on LEGO brick colors, consult *http://www.peeron.com/inv/colors*.

TRANSLUCENT   A semitransparent quality such that light can pass through but detailed shapes are not visible. Translucent bricks are made in many colors.

VIEWING ANGLE   The arc, measured in degrees, at which an LED package is designed to project light.

VINYL CUTTER   A machine designed to cut rolls of adhesive-backed vinyl into text, logos, and other designs.

WINDOW LIGHTING   The patterns of light projecting outward through a building's windows that define the building's contour and character.

WOODLAND SCENICS   A manufacturer of scenery and lighting products for the model railroad market that can also be useful for lighting LEGO builds (*https://www.woodlandscenics.com*).

WORLD OF LIGHTS (WOL)   An event at Brickworld where the room lights are reduced or turned off entirely for a period of time, allowing attendees to view the lights built into the MOCs on display. Some other LEGO fan conventions have similar reduced lighting events.

PHOTO CREDITS

All photographs are courtesy of the author, except for those listed below.

Airport. Photo © **CHRIS ROSEK** / *Flickr.com*.

The Batcave. Photo © **WAYNE HUSSEY & CARLYLE LIVINGSTON II**.

Bio Lab One. Photo © **WAMI DELTHORN** / *Flickr.com*.

Celebricktion. Photo © **HARRY & AUSTIN NIJENKAMP**.

Coffee Shop. Photo © **FOOLISH BRICKS** / *Foolishbricks.com*.

Crystalline Tree. Photo © **DUNCAN LINDBO & AUTHOR** / *Flickr.com*.

Cyberpocalypse. Photo © **CHRIS EDWARDS** / *Flickr.com*.

The Dragon's Wrath. Photo © **HOLLY WACHTMAN** / *Flickr.com*.

Ford E-450 Ambulance. Photo © **MICHAEL GALE**.

Ghost Ship. Photo © **MATT DE LANOY** / *Flickr.com*.

Henley Street Bridge. Photo © **BEYOND THE BRICK** / *Twitter.com*.

Hogwarts. Photo © **HYUNGMIN PARK** / *Flickr.com*.

Hope Castle. Photo © **PATTI RAU** / *Flickr.com*.

Hydroponics Research & Development Facility. Photo © **JON BLACKFORD** / *Flickr.com*.

Ice Sculptures. Photo © **CHAIRUDO** / *Flickr.com*.

Imagine. Photo © **SHARPSPEED & AUTHOR** / *Flickr.com*.

Imperial Star Destroyer Chimaera. Photo © **JERAC** / *Flickr.com*.

In a Good Book. Photo © **PATTI RAU** / *Flickr.com*.

In a Good Book, close-up detail. Photo © **TONY SAVA**.

Jenn's Fish aka My First MOC Evah!!. Photo © **JOE MENO** / *Flickr.com*.

The Kent Theatre. Photo © **JARREN HARKEMA & AUTHOR**.

Kubo and the Two Strings. Photo © **BEYOND THE BRICK** / *Twitter.com*.

Library of Druidham. Photo © **BENJAMIN STENLUND** / *Flickr.com*.

Magic Angle Sculpture. Photo © **JOHN V. MUNTEAN**.

Millennium Falcon. Photo © **MARSHAL BANANA**.

National Mall. Photo © **WAYNE TYLER**.

Octan Park. Photo © **RYAN DEGENER**.

Opee Sea Killer. Photo © **MATT DE LANOY** / *Flickr.com*.

Peterbilt Quint Axle Dump Truck. Photo © **DENNIS GLAASKER**.

Prelit minifig accessories. Photo © **BRICKSTUFF**.

A River Runs Through It. Photo © **BARBARA HOEL** / *Flickr.com*.

Sandcrawler. Photo © **MARSHAL BANANA**.

Speeder Repairs. Photo © **BENJAMIN STENLUND** / *Flickr.com*.

Stained Glass Windows. Photo © **BILL WARD & AUTHOR** / *Flickr.com*.

A street lamp. Photo © **BRICKSTUFF**.

The Tour Continues. Photo © **BILL TOENJES** / *Flickr.com*.

Tree of Worlds. Photo © **HOLLY WACHTMAN** / *Flickr.com*.

Ulric the Firebright. Photo © **CORNBUILDER** / *Flickr.com*.

Union Pacific EMD SD70 Ace Locomotive. Photo © **DENNIS GLAASKER**.

United States Capitol. Photo © **WAYNE TYLER**.

Villa Amanzi. Photo © **ROBERT TURNER**.

Wiring diagram for truck lighting design. Photo © **BRICKSTUFF**.

World of Lights, Brickworld Chicago 2009. Photo © **JOE MENO**.

World of Lights, Brickworld Chicago 2011. Photo © **BILL TOENJES** / *Flickr.com*.

LIST OF MOCS

INDEX

RESOURCES

Visit *https://nostarch.com/lego-lighting-book* for errata and more information.

More no-nonsense books from  **NO STARCH PRESS**

THE LEGO NEIGHBORHOOD BOOK
Build Your Own Town!

BY BRIAN LYLES *AND* JASON LYLES
204 PP., $19.95
ISBN 978-1-59327-571-6

THE LEGO NEIGHBORHOOD BOOK 2
Build Your Own Town!

BY BRIAN LYLES *AND* JASON LYLES
192 PP., $19.95
ISBN 978-1-59327-930-1

THE LEGO ARCHITECTURE IDEA BOOK
1001 Ideas for Brickwork, Siding, Windows, Columns, Roofing, and Much, Much More

BY ALICE FINCH
232 PP., $24.95
ISBN 978-1-59327-821-2

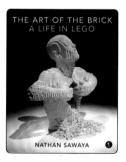

THE ART OF THE BRICK
A Life in LEGO

BY NATHAN SAWAYA
248 PP., $39.99
ISBN 978-1-59327-588-4

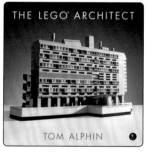

THE LEGO ARCHITECT

BY TOM ALPHIN
192 PP., $24.99
ISBN 978-1-59327-613-3

LEGO TRAIN PROJECTS
7 Creative Models

BY CHARLES PRITCHETT
208 PP., $24.95
ISBN 978-1-7185-0048-8

PHONE:
800.420.7240 OR
415.863.9900

EMAIL:
sales@nostarch.com
WEB:
www.nostarch.com